MANAGING THE MACHINE

How to Lead, Adapt, and Thrive in the Age of Artificial Intelligence

WHITE PINE PUBLISHING

This publication is designed to provide accurate and authoritative information in regard to the subject matter covered. It is sold with the understanding that neither the author nor the publisher is engaged in rendering legal, investment, accounting or other professional services. While the publisher and author have used their best efforts in preparing this book, they make no representations or warranties with respect to the accuracy or completeness of the contents of this book and specifically disclaim any implied warranties of merchantability or fitness for a particular purpose. No warranty may be created or extended by sales representatives or written sales materials. The advice and strategies contained herein may not be suitable for your situation. You should consult with a professional when appropriate. Neither the publisher nor the author shall be liable for any loss of profit or any other commercial damages, including but not limited to special, incidental, consequential, personal, or other damages.

Book Cover by White Pine Publishing LLC via Canva

Contents

Introduction

It's Monday morning, and you're staring at your calendar. Another meeting notification pings for your attention. Your team is spread across kitchen tables and office cubicles, juggling video calls, Slack threads, and whatever communication tool your company has settled on this quarter. You're expected to deliver results, guide people through change, and stay the steady hand while new technology reshapes everything from project planning to performance reviews. Somewhere between coaching an employee who wonders if a chatbot will take their job and wrestling with an "ethical approval workflow" nobody quite trusts, you realize something: leading right now isn't covered in any management playbook you've read before.

That's why this book exists. It's a field guide for managers, directors, VPs, and executives steering teams through digital change and the unpredictable rollout of artificial intelligence at work. Inside, you'll find frameworks for introducing new technology, ways to keep your team steady under pressure, and language for the harder conversations about ethics and trust. The aim isn't just to help you keep up. It's to help you build a culture where technology backs up your people instead of crowding them out.

The anxieties are real, and we're not going to pretend otherwise. You might worry that automation threatens jobs, including your own. You might worry your team will fall behind as tools change faster than anyone can learn them. The ethical questions can feel murky, and resistance often shows up the moment someone says "AI" out loud. If a phrase like "machine learning" makes your eyes glaze over, you're in good company. This book doesn't assume you're a data scientist or a true believer in the tech. It meets you where you are, in plain language, with stories that make room for both hope and hesitation.

This isn't a book you just read and shelve. You'll find reflection prompts, things to try with your team, and room to stay skeptical while you think it through. The tone is conversational, the outlook is realistic, and we've tried hard to keep the buzzwords out. By the end, the goal is for uncertainty to feel less like a threat and more like something you know how to work with.

Getting the balance right between smart technology and genuine judgment isn't just possible. It's the whole point. The best teams ahead will be the ones that use new tools without losing what actually makes a team work: trust, judgment, and care for each other. The goal here was never to replace people with machines. It's to help you turn these tools into something that makes your team's work better.

So, welcome. Think of this as an invitation, not a lecture, a place to bring your curiosity and your skepticism in equal measure.

Along the way you'll also find a link to a short companion tool, the Decision Line, a quick diagnostic that shows where your own delegation and transparency habits currently sit. More on that at the end of the book. For now, let's get into it.

One

The New Leadership Landscape: AI in the Workplace

Artificial intelligence is not a substitute for human intelligence; it is a tool to amplify human creativity and ingenuity.

Fei-Fei Li

PICTURE AI-POWERED ASSISTANTS THAT spot staffing gaps or workflow bottlenecks before you do, sifting through data as it comes in and helping you decide faster and with more confidence. They schedule the right people at the right time, flag trends worth watching, and suggest a next step, quietly becoming part of how leadership actually happens day to day. That's a real shift: it invites managers to partner with the technology instead of just managing tasks around it.

Organizations that welcome AI early aren't just keeping up. They're the ones setting the terms other companies end up following. The tools unlock agility and new ways to solve old problems, and this chapter looks at how AI is reshaping leadership roles and the day-to-day rhythm of work.

The Age of AI: Shaping Tomorrow's Work

Imagine a colleague who never sleeps and gets faster at spotting patterns the more experience it has. That's roughly what AI is: technology that solves problems the way people do, but at a speed no human matches. Instead of relying on memory or gut instinct, it works through large amounts of information and offers a first read, which is often useful and sometimes wrong. Traditional software follows fixed rules. AI adapts and improves the more it's used, the way an employee gets better the longer they're on the job.

At the core of this is machine learning: computer systems that learn from data instead of being told exactly what to do. A chatbot in your HR system might only answer basic questions today, but it can get better at picking up nuance and intent as the vendor improves it over time. Some AI tools scan thousands of resumes in seconds and flag the strongest matches, something no human recruiter could do at that speed. In manufacturing, predictive maintenance tools catch small shifts in equipment performance early enough to prevent an expensive breakdown.

Leaders run into AI constantly without noticing it. Smart assistants schedule meetings around everyone's availability so calendars stop colliding. Marketing teams use recommendation engines to personalize outreach and match offers to what a buyer actually wants. Finance teams use algorithms to flag anomalies that might signal fraud, backing up the audits that already exist.

None of this is science fiction. It's already woven into daily operations across most sectors, quietly supporting decisions and workflow design (Byrne, 2023).

Embracing AI does more than smooth out operations. Early adopters make faster, better-supported decisions and get ahead of market trends before competitors notice them. Automating repetitive work frees people up for the creative work that actually grows the business. Companies that hesitate pay for it: their teams keep chasing information and doing redundant work while competitors move faster. In industries that shift quickly, slow adaptation can threaten more than market position. It can threaten the company's survival (Mayer et al., 2025).

Leaders also face real concerns about AI's trustworthiness and its effects on their teams. Employees worry about accuracy, cybersecurity, shifting job requirements, and whether AI-driven decisions can be trusted. The same surveys that surface those worries also show most workers expect their own company to roll out AI responsibly, an optimism that depends on leadership actually making good, transparent choices (Mayer et al., 2025). Handled well, AI integration builds trust instead of eroding it.

AI also opens up ways to develop talent. Platforms can spot skills gaps and recommend personalized learning for each team member, helping companies upskill people as business needs change. On cross-functional projects, AI can surface strengths within a team and suggest pairings nobody would have thought to try, like a marketing manager and a product designer teamed up because the system noticed their recent work overlapped in a useful way. It ends up acting like both an analyst and a matchmaker for talent (Byrne, 2023; Mayer et al., 2025).

The real promise here isn't machines replacing people. It's people getting more done because intelligent systems are backing them up. Some call this "collaborative intelligence," human judgment paired with AI's speed and precision. In a crisis, a leader might combine experience-based instinct with AI-run simulations to weigh options and forecast outcomes fast, landing on a decision informed by both. Treating AI as a partner instead of just a tool lets organizations move faster and adapt better to change (Mayer et al., 2025).

That foundation, what AI can and can't do, is what makes the next question worth asking: how is the role of a people leader actually changing because of it?

The Evolving Role of People Leaders

The workplace is changing fast, and leadership with it. Leaders used to supervise tasks and check compliance. Now, as AI works its way into daily operations, people leaders are becoming more like architects, designing environments where human skill and machine intelligence work together (Stanin, 2025). Think less supervisor checking boxes, more conductor bringing together musicians and instruments that weren't built to play together on their own. The job is making sure each player, human or algorithm, contributes what it's actually good at.

Managers today are doing less overseeing and more orchestrating: facilitating instead of dictating, building workflows where humans and AI play complementary roles (Stanin, 2025). Picture a marketing manager who redesigns how campaigns get planned. AI tools sift through customer data and flag trends fast; the team turns those trends into stories and strategy the AI can't write on its own. Instead of micromanaging, the manager builds feedback loops between people and the tools, and makes room for experimentation across teams.

Making that work means leaders get comfortable pulling on both data and human judgment. During a quarterly review, AI might flag a sales dip in a specific region, but it's the team's understanding of the local market that explains why, maybe a competitor ran an unexpected promotion. Good managers build in a step where employees can check algorithmic findings, push back on AI recommendations, and work out solutions together that neither side would land on alone (Frimpong, 2025; Stanin, 2025).

It helps to know where each side is actually stronger. AI can process huge datasets, automate repetitive paperwork, and spot correlations a person would never catch on their own. But intuition, empathy, ethical judgment, and creativity are still things only people bring, especially in ambiguous or culturally complex situations (Frimpong, 2025). In customer service, AI might predict which clients are likely to churn, but it's the agent who hears the distress in someone's voice and handles the delicate work of rebuilding trust. A leader's job is knowing when to trust the AI's speed and when the moment calls for a person to slow down and read the room.

Getting that balance right means teaching teams to question AI tools, not just use them. The best leaders make skepticism a normal part of the culture instead of something people apologize for. They set up regular time to talk openly about where automation is working and where a person had to step in and fix something (Stanin, 2025; Frimpong, 2025). Sharing specific examples, like "AI flagged the wrong trend because it missed a holiday spike" or "our designer caught a shift in client mood that changed our whole pitch," keeps critical thinking part of the conversation as the tools get adopted.

Things keep changing, and leaders don't get through that by having all the answers. They get through it by modeling curiosity and staying steady when plans don't hold. Picture a mid-size logistics firm that rolls out a new scheduling AI, and the tool unexpectedly reshuffles delivery routes. Glitches cause delays for a week. The team's leader runs a fast retrospective: what worked, what didn't, who adapted and how. Running small experiments, rotating responsibilities, and calling out quick recoveries builds a team that treats setbacks as something to work through instead of something to fear (Stanin, 2025; Frimpong, 2025).

Nothing replaces actually talking to people. AI can boost efficiency, but it can also leave team members feeling unsure of where they still add value. Good leaders address that directly: they run town halls to explain what's coming, admit their own uncertainty out loud, and talk about both the collaborations that worked and the ones that didn't. They check in with people one on one and make psychological safety a real practice, not just a phrase on a slide (Stanin, 2025). Peer mentoring or informal support circles help reassure people that their expertise still matters.

Clear communication matters here: explain why a new AI tool exists, share real examples of it helping, and be honest about what's still being adjusted. A leader who says, "this isn't about replacing anyone, it's about making space for your skills to matter more," builds trust in a way people notice. Recognizing collaborative behavior and staying open to feedback on AI decisions keeps that trust intact over time. As algorithms start influencing more choices, ethical oversight matters too: leaders need to name bias when they see it and make sure AI-driven decisions still reflect the organization's actual values.

With a clearer sense of how the leader's role is shifting, it's worth clearing up a few myths that tend to get in the way.

AI: Myths and Realities for Leaders

Mid- and senior-level leaders are the ones bridging technology and people day to day. Part of the job is adopting new tools; the bigger part is steering and protecting the team using them. As that role shifts, a lot of managers run into the same obstacle: persistent myths about AI that stir up anxiety right where trust and judgment matter most. Clearing those up opens the door to moving with more confidence.

Myth 1: AI Will Replace All Jobs, Including Team Leaders and L&D Roles

It's easy to imagine a future where machines handle everything and teams become obsolete. Headlines about automation and AI beating humans at chess, medical diagnostics, or content generation feed that fear. A lot of people worry that learning departments, team leads, and other people-focused roles will disappear as AI gets smarter and faster (Goswami, 2024).

The myth sticks around because of media hype and because people have watched digitization happen before. Stories about job loss circulate in break rooms and video calls. When teams see a bot take over scheduling or training reminders, it's easy to assume bigger replacements are next. But that overlooks how these things actually play out. Automation is genuinely better than people at routine data processing. It's still weak on judgment, emotional context, and coaching. AI can recommend personalized learning modules, but it's the team leader who brings the context, motivation, and connection that actually makes people grow.

Leaders who name this myth directly can audit their own workflows and see which tasks are actually ready for automation and which still need a human touch. That reduces anxiety, and it frees teams to focus on the work only people can do, like supporting someone's growth or managing a complicated relationship (Goswami, 2024).

Myth 2: Only Tech Experts Can Lead with AI

Another myth is that real "AI-ready" leadership requires deep technical skill, coding ability, or a machine learning background. Leaders without a technical background often hold back from AI initiatives, assuming they don't have the expertise to shape outcomes. That sidelines some of an organization's best communicators and strategists from a conversation they belong in.

This myth survives because AI conversations default to jargon and case studies full of specialists and data scientists. But strong leadership was never about mastering every tool. It's about pulling resources together, asking good questions, and making informed calls. Picture a project leader working with a data analytics team: the analysts build the model that predicts customer churn, and the leader turns that into action, shapes how it gets rolled out, and makes sure it actually serves the business goal.

Managers who stay curious and build real partnerships with technical experts open the door to both innovation and inclusion. They don't need to write code, the same way a coach doesn't need to play every position, but they do need the confidence to bring people together, question assumptions, and keep ethics part of the conversation (Goswami, 2025). One concrete step: set up regular sessions where business and technical teams actually talk to each other.

Fact 1: AI Supports, Not Replaces, Decision-Making

AI works more like a GPS than an autopilot: it offers routes and insight, but the leader still decides where to go. Even where predictive analytics reshape a process, the final call still needs human judgment (Goswami, 2025). A manager looking at AI-driven performance dashboards during quarterly reviews still relies on observation, empathy, and frontline feedback before deciding on a promotion. The AI can't see the quiet signs of potential or the personal circumstances that affect someone's numbers.

Managers who get this use AI to inform decisions, catch blind spots, and surface opportunities, not to make the call for them. Building a habit of discussing AI recommendations alongside the qualitative, human read on a situation keeps both the results and people's trust intact.

Fact 2: Human Values Are Essential for Successful AI Integration

Successful AI adoption depends on keeping human values, ethics, fairness, accountability, built into every stage of the decision. Leaders who treat AI output as objective truth, or who hand off responsibility to the algorithm, risk real missteps: bias creeps in and trust erodes (Goswami, 2025). Relying only on an AI hiring tool, for instance, can amplify discrimination if nobody checks for biased patterns or considers context.

Leaders need to keep processes transparent, run regular bias audits, and stay accountable for how AI shapes decisions about people. A concrete step is writing clear guidelines for evaluating AI suggestions, weighing both the benefit and the ethical cost. That protects individuals, and it protects the organization's credibility.

With a clearer view of what AI can and can't do, the next question is what mindset actually makes use of that. Seeing past the hype toward AI's real promise and its real limits is what makes the next set of habits, openness, curiosity, and empathy, worth building.

Building Your AI Mindset

With those myths out of the way, it's worth focusing on the mindset that makes people-centered success last. That starts with how leaders approach AI in the first place: not just as a technical problem, but as a chance to shape the organization's culture on purpose.

Openness, Curiosity, and Collaboration

An AI mindset starts with openness: being willing to explore beyond the usual routines. Picture a manager who notices their team feeling uneasy when a new analytics tool launches. Instead of dictating how to adopt it, they open a weekly "explore and learn" hour where people bring questions and even their frustrations. Employees swap tricks they've found or pitch creative uses, and that kind of exchange often turns up ways to make everyday work smoother (Palomino, 2025). Openness tends to spread. Curiosity follows when a leader is genuinely enthusiastic about learning rather than performing it. Experimentation doesn't need to be a moonshot: trying an internal chatbot for tracking PTO requests and then gathering feedback in a retrospective works fine. Small tests like that lower the stakes and show people it's safe to try something, see how it goes, and adjust. Celebrated experimentation speeds up learning and makes AI adoption feel less like a chore (Palomino, 2025).

Curiosity Over Certainty

Curiosity does more than make uncertainty tolerable. It surfaces what leaders don't yet know. Picture a regional director rolling out a scheduling algorithm who doesn't assume the tool is flawless. She holds an open Q&A and lets frontline employees ask "what if" questions: what happens if the model misreads my time-off needs, how does it handle exceptions. Encouraging those skeptical questions surfaces blind spots that, left alone, turn into bias or a morale problem. Teams that keep asking questions tend to catch mistakes early and spot opportunities that more rigid planners miss (Palomino, 2025). A question-driven culture means regularly asking what assumptions are being made, whose voice is missing, and what happens if something goes wrong. A little humility paired with consistent inquiry goes a long way ("AI Is Not the Silver Bullet," 2025).

Empathy and Technology Working Together

Putting people first matters throughout any AI rollout. Compassionate leadership isn't just patience, it's actively seeking out concerns before moving ahead. A senior leader might run listening sessions before introducing workflow automation, giving employees room to say honestly that they're afraid of losing their job or of the process feeling less human. The leader explains how automation frees the team for more meaningful customer work, and asks for input on designing workflows that add to human work instead of replacing it (Palomino, 2025). Anonymous surveys give people another channel to raise concerns, which lets leadership adjust course or add support where it's needed. When empathy guides these decisions, morale holds up better and resistance tends to drop ("AI Is Not the Silver Bullet," 2025).

Practical First Steps for Leaders

The most effective way to build this mindset is starting small and visible. Pick one low-risk pilot, maybe a scheduling assistant or a sentiment analysis dashboard, and design it with your team rather than for them. Be clear that it's an experiment, not a permanent change, and get everyone, including the skeptics, involved in testing it and giving feedback. Celebrate progress as it happens, however small, so the wins feel shared (Palomino, 2025). Keep communication honest: update the team regularly, admit what isn't working, and share what you're learning. Practices like that chip away at fear and build the capability both leaders and teams need for bigger initiatives later.

None of this stops once you've run a pilot. Fostering openness, asking good questions, leading with empathy, and running practical pilots all build on each other over time, and they're what prepare managers and teams to keep up as the technology keeps changing. Building an AI mindset isn't a box to check off. It's an ongoing way of keeping people at the center of every change (Palomino, 2025; "AI Is Not the Silver Bullet," 2025).

Wrapping Up

This chapter looked at how AI has gone from distant possibility to active partner in how organizations actually work, from chatbots handling daily tasks to predictive tools catching trends early. Leaders who lean into that find new ways to innovate and keep their teams focused on the human, creative work that actually delivers value.

Embracing AI doesn't mean losing your leadership touch or your team's strengths. It means becoming more like a conductor, blending technology and people to get the best result. Staying open, asking questions, and keeping empathy in the decision are what let you guide a team through this kind of change. AI can amplify what's possible, but it doesn't replace real leadership. That's still on you.

Two

Understanding the Human-AI Partnership

> *The main business of humanity is to do a good job of being human beings ... not to serve as appendages to machines, institutions, and systems.*
>
> Kurt Vonnegut (Player Piano)

IMAGINE A MARKETING DIRECTOR at a mid-sized company who turns to an AI-powered data analytics platform, hoping it will reveal the next big customer trend and guide her team's strategy. The technology quickly processes mountains of data, highlighting emerging patterns that promise competitive advantage. But here's the hitch: her team's gut feelings, cultural knowledge, and creative instincts don't always line up neatly with what the AI suggests. They debate whether to follow the algorithm's lead or trust their own experience, trying to find common ground without losing momentum. That push and pull is the real-world challenge most leaders eventually face: getting human insight and artificial intelligence to work together instead of pulling in opposite directions.

Collaborative Intelligence Defined

Chapter 1 looked at how the role of the people leader is changing in an AI workplace, and why balancing human and machine strengths matters. This chapter gets more specific: what does that balance actually look like in practice, and how do you build it into the way your team works day to day?

Collaborative intelligence is what you get when you deliberately pair human strengths, creativity, contextual understanding, ethical judgment, intuition, with the speed and scale AI brings to processing information. It's less about picking a winner between people and machines and more about designing teams where each side does what it's actually good at. AI can drive the data processing and generate the first set of recommendations fast. Humans do the steering: deciding which of those recommendations actually make sense given what they know about the customer, the market, or the moment.

Take that same marketing director. Her analytics platform sifts through consumer behavior data and flags emerging trends and campaign ideas. Her team layers in market intuition and cultural awareness to interpret those findings, adjust the messaging, and connect the product to what customers actually need. The AI's output becomes fuel for sharper questions and faster pivots, not a verdict the team has to accept as-is.

The same pattern shows up in executive boardrooms. An AI might flag underperforming product lines across several regions, but it takes someone who knows the business to figure out whether that's economic conditions, cultural differences, or a competitor's move, and to craft a response that fits. Some organizations build cross-functional teams around this idea: AI automates the repetitive financial analysis, freeing finance, HR, and operations to spend their time on scenario planning and risk decisions that actually need judgment.

The strength of collaborative intelligence is that it multiplies what a team can do. Human reasoning alone is limited by attention, memory, and bias. AI alone misses judgment, ethics, and the ability to adapt to something it hasn't seen before. Put the two together well and teams get faster at spotting risk, catching opportunity, and responding to change. AI extends how much a team can take in. People give that information meaning, weighed against company values and what customers actually need.

Customer service is a good example of this working at scale. AI chatbots field the routine, high-volume questions around the clock and hand off anything complicated or emotionally charged to a person who can read the situation and respond with real empathy. Response times improve, workload pressure drops, and customers still get treated like people. Human agents also review what the AI suggests, catch its mistakes, and flag issues, which is part of what makes the system better over time.

None of this happens automatically. Teams that don't trust the AI's recommendations tend to ignore them, especially when they can't see how the algorithm got there. Without clarity on who decides what and when, things stall or slip through the cracks. And technical staff, business users, and leadership often have different assumptions about what the AI is actually for, sometimes one group treats its output as gospel while another won't touch it at all.

The fix is mostly about visibility: explain how the AI reaches its conclusions, make it explicit who owns which decisions, and build in regular feedback between people and the system. Leaders who do this well find that transparency builds trust, and that naming these things out loud is what lets a team sharpen both its use of the tool and its own judgment.

Redesigning Workflows for Synergy

Chapter 1 established that balancing human and machine strengths isn't optional anymore. The next question is more practical: how do you actually restructure the way your team works so that people and AI operate side by side without burning anyone out (Amane Hoshimi, 2025; Eastwood, 2025)?

Start by mapping your current workflow with fresh eyes. Walk through every step of a process you care about, onboarding, campaign management, product development, whatever it is, and mark what's repetitive, rules-based, and data-heavy. Logging sales calls, generating routine reports, that kind of thing is a good candidate for AI to take over. Then flag the moments that need judgment, creative thinking, or context your team actually has. A manager doing this might realize that updating a social calendar can be automated, but responding to a trending customer complaint still needs a person. Mapping it out this way keeps AI from quietly swallowing work that only a person should be doing (Amane Hoshimi, 2025).

If you want a quick read on where your own delegation habits actually land, that's what the Decision Line was built for, a short scenario-based tool at decisionline.whitepine.pub that scores how you delegate and how transparent you are about it. Ten questions, no email required to see your result.

Once you know what's on the table, resist automating everything just because a vendor makes it sound easy. Some situations need a human's heart or ethical judgment, and no AI has that yet (Eastwood, 2025). An AI chatbot responding to a customer who just lost someone close to them, without empathy, can go sideways fast and cost real trust. Keep people in charge of hiring, performance reviews, and sensitive conflict resolution, and let the bots handle appointment reminders and invoicing. This isn't about resisting efficiency. It's about protecting the parts of the job that actually need a person.

As AI takes on more of the routine work, leadership needs to update roles, not just tools. Job descriptions should start reflecting skills like interpreting algorithmic output, working alongside AI on a shared project, and improvising when something doesn't go as planned. A sales leader doesn't need to compete with a lead-scoring tool. They can combine its output with what they pick up from a client's tone on a call. The goal isn't for people to keep pace with machines. It's using the machine to free up time for the work that actually needs a person: creativity, hard problems, and building trust (Amane Hoshimi, 2025).

Big rollouts tend to stumble. Small, focused pilots tend to work. Pick something contained, like event scheduling or an internal help desk, run a short pilot, and gather feedback from everyone touching the new system, not just supervisors. Weekly check-ins surface the friction points and the surprises early. A team piloting an AI meeting scheduler might adjust its settings after a week of real conflicts show up that the tool didn't anticipate. Treat these pilots as something you keep adjusting, not something you launch and walk away from (Eastwood, 2025).

Small wins add up. Skepticism turns into curiosity, and teams start asking what else could free up their time for the work that matters most.

Balancing Efficiency and Empathy

Once you know where AI genuinely helps, the question shifts. It's no longer just what to automate, it's how automation can make you a better, more present leader rather than a more distant one. Letting machines handle the repetitive stuff means you get more time for mentoring, coaching, and the harder conversations, instead of getting buried in admin. A manager who used to spend hours a week scheduling one-on-ones and approving leave requests gets that time back with a scheduling assistant, and can actually show up for the moments that matter (Empathy: What It Means for an AI-Driven Organization, 2025).

AI is good at surfacing patterns. It can't tell you what they mean. If a system flags a dip in someone's performance or rising stress signals on a team, that's a starting point, not an answer. It takes a manager's own read on the situation to know whether someone's dealing with a sick family member, burning out, or just swamped this month (Eklund, 2024). AI's real value here is helping you catch these signs earlier, not replacing your own judgment about what to do next. The best leaders still sit down and ask, and actually listen to what comes back, instead of trusting the dashboard to tell the whole story.

Getting this right takes intention. Dashboards and AI tools are fine for operational oversight, but feedback, career conversations, and anything personal still need a real conversation. Decide in advance where the human moments have to happen, video call, phone, in person, and protect them. Automate the weekly status update if you want, but keep the monthly coaching session live. When someone brings you a growth question or something sensitive, showing up in person says something an algorithm can't (Empathy: What It Means for an AI-Driven Organization, 2025; Eklund, 2024).

Getting a team ready for this kind of partnership takes real investment. Short workshops on emotional intelligence, resilience, and using AI responsibly go a long way toward taking the mystery out of the technology while keeping respect and connection at the center. Some companies name a few "AI ambassadors" on each team, people who model good use of the tools and field questions from peers. Regular forums where people can talk honestly about what's working and what isn't tell the team that asking questions is welcome, not a liability. All of this helps push back against the fear that people are being quietly replaced (Empathy: What It Means for an AI-Driven Organization, 2025).

None of this is separate from the bottom line. Pairing AI-driven efficiency with real care for people isn't just good culture, it shows up downstream in recruitment, retention, and how well the organization holds up under pressure. Every task you hand off to automation is worth pairing with a deliberate investment somewhere else: a better conversation, real recognition, a shared sense of purpose.

A few concrete places to start: look at what's actually being helped by automation right now versus what still needs your full attention. Use whatever time automation frees up and reinvest it in the relationships on your team. Check in regularly on whether the AI tools are actually helping morale and engagement, not just throughput, and adjust based on what people tell you. Bring your team into designing the process instead of just handing them the result. And model the transparency and empathy you want to see, so people know efficiency and humanity aren't competing goals here.

Real-World Applications of Human-AI Teams

HR departments know the pressure of a hiring surge better than most: mountains of resumes, burned-out recruiters. AI-enabled recruitment doesn't replace that team, it works alongside it. Think of AI as a fast first pass, screening for skill matches and flagging strong candidates for a recruiter to check. Masking demographic details in that first screen can reduce hidden bias. But the final call, whether someone actually fits the team and the company, is still deeply human work. Managers read the cues an algorithm can't, and they're the ones who build a new hire's sense of belonging from day one. Handled this way, hiring gets both more inclusive and more likely to stick (Masood, 2025).

Performance reviews often feel arbitrary or biased, and that disengages people fast. AI can help here too, catching patterns like a sudden drop in output or a shift in team dynamics before they become a bigger problem. But only a manager has the context to know what a signal actually means. If an AI report shows a team member's productivity has dropped, the useful move isn't to correct them on the spot, it's to ask what's going on. Maybe it's a family situation, maybe the project scope shifted under them. That context is what turns a generic check-in into a conversation that actually helps someone grow (Mayer et al., 2025; Masood, 2025).

Customers expect personalization now, and businesses that can't deliver it start losing ground. AI-driven personalization tools can process browsing history and purchase patterns fast enough to tailor offers in real time. But automated hospitality only goes so far. Frontline staff still need to notice when the bot's hit its limit. A retail associate watching a chat log might see the bot handle a shipping question fine, then catch frustration in the customer's tone and step in with a live call. That handoff, AI on the routine stuff and a person on the moment that actually needs care, is what keeps customers loyal (Mayer et al., 2025; Masood, 2025).

Sales teams face something similar with the sheer volume of prospects to track. An AI-powered CRM can surface promising leads buried in the data that a person would likely miss. But closing still comes down to old-fashioned rapport. A salesperson reads hesitation, navigates a tricky conversation, and rebuilds trust with a client who's gone quiet, work an algorithm can flag but can't do. The AI points at the opportunity. The person still has to close it.

Bringing It All Together

Blending human insight with AI recommendations isn't about picking a side. It's about building teams where both play to their strengths. Collaborative intelligence works when people bring empathy, context, and creativity, and AI brings speed and scale. Redesigning workflows deliberately, communicating clearly, and being explicit about who owns which decisions is what makes a team more adaptable, not just more efficient. The leaders who get the most out of this don't automate for its own sake. They use the technology to support judgment, trust, and the kind of connection that actually holds a team together.

For a mid-sized organization working through this, there's no shortcut. Map your processes, build in real feedback loops, and remember that the good outcomes come from combining data with actual human experience. Get the balance between efficiency and empathy right, and your team is better positioned to handle change, catch new opportunities, and hold onto the relationships that matter, inside the company and with your customers. Stay curious, keep asking questions, and keep the conversation going. That's what makes AI work for your team instead of the other way around.

Three
The Empathetic AI Leader

Seek first to understand, then to be understood.

Stephen Covey

IT'S THE MIDDLE OF a project launch, deadlines stacking up, notifications firing every few minutes, when a manager pauses a group call to check on a remote team member. Do they have what they need? Are they underwater? After hearing the answer, the manager acts on it, rebalancing a task or two, or just making sure the person knows they're seen. That small pause is the whole chapter in miniature: empathy isn't disappearing as AI takes on more of the work, it's just changing shape, and it takes some deliberate habits to keep it real when so much of the day happens through a screen.

Defining Empathy in the Age of AI

Empathy at work used to mean noticing when someone was having a hard day. Now, especially for leaders managing hybrid teams, it means actively trying to understand what a colleague is going through even when you only ever see them through a screen. That means listening for what isn't said, noticing when someone goes quiet on a call, and following up with an action, not just a kind word. The manager pausing that call to check on a struggling teammate, and then actually rebalancing the workload afterward, is doing both halves of it: noticing, and responding (Uy, 2024).

Digital environments blur or block a lot of the cues that face-to-face conversation gives you for free. In an office, body language fills in a lot of gaps. Online, that's mostly gone. Sustaining real empathy remotely takes more deliberate effort: turning on video for routine check-ins so you can actually see someone's face, mixing up how you communicate (email, chat, a quick voice note) so nuance doesn't get flattened, or building in something informal like a virtual coffee chat. Even something as small as asking people to rate their mood on a scale at the start of a meeting opens a low-pressure channel for honesty (Uy, 2024).

Fostering empathy remotely isn't easy, and it's worth taking seriously. Technology that's overused or applied without much thought can quietly turn people into task-completing units instead of, well, people. Leaders who pair digital tools with real emotional awareness build cultures where people feel valued for who they are, not just what they produce. Pulse surveys are a decent quick read on team sentiment, and a targeted follow-up conversation catches what the survey misses. Small rituals like a virtual coffee catch-up or a skill-sharing session build the kind of rapport that's harder to come by when nobody's in the same room (Uy, 2024).

Empathy isn't a soft add-on here, it's the foundation trust actually gets built on, especially when a new technology or workflow is rolling out. The leaders who do this well don't start with a technical demo. They start by naming the uncertainty their team is probably feeling, inviting people to say what worries them, listening without multitasking, and actually using what they hear to shape how the rollout happens. Leading with that kind of empathy earns trust before the change even lands, and that trust shows up later in engagement and performance (Lewis, 2021; Uy, 2024).

None of this requires reinventing how you communicate, just treating every interaction, even a short one, as a chance to actually connect. Looking at the camera instead of the screen, muting notifications, resisting the pull to multitask during a call, these are small things that change how present you seem. A five-minute video call often does more for someone's morale than three clipped emails, especially when things are stressful. None of these are innate skills. Anyone can build them with practice (Lewis, 2021).

Empathy also underlies good coaching. Trading directive instructions for open questions, real listening, and letting people help shape their own action plan builds the kind of trust that makes a team resilient. Managers who admit their own mistakes, celebrate wins honestly, and show up consistently set the tone for a workplace where people can bring their actual selves to work without worrying about being judged for it (Lewis, 2021).

Chapter 4 picks up from here and looks at how ethical AI leadership grows out of this same ability to understand what people actually need underneath a technology change. As organizations keep moving through digital transformation, empathy stops being a nice-to-have. It's what keeps the technical progress lined up with what the organization actually values, and with the people living through the change (Uy, 2024; Lewis, 2021).

Leading with Emotional Intelligence

Emotional intelligence is what turns empathy into something practical for leaders running AI-powered teams day to day. As automation and data reshape more of the work, leaders who can read and manage emotions, their own and their team's, are the ones who connect human values to actual technology adoption (Dwivedi, 2025; Ramakrishnan & Krupskyi, 2025). Even when an algorithm is helping make the call, the ability to connect with people and keep them motivated still matters most.

Self-Awareness in Decision-Making

Self-awareness sharpens judgment, especially when AI is putting recommendations in front of you. It means noticing your own habits and how your preferences shape your reaction to a new tool. Say you get an AI-generated report ranking team performance, and you lean heavily toward traditional productivity metrics. You might dismiss recommendations that challenge your comfort zone without even noticing you're doing it. A manager who catches that discomfort in themselves, instead of acting on it right away, might bring the team together to talk through the data and where it's coming from. That conversation surfaces blind spots and grounds the decision in both the data and the team's actual experience. Practicing this kind of self-awareness builds the habit of questioning what assumptions are baked into the AI's output in the first place (Dwivedi, 2025).

Self-Regulation for Calm Adaptation

Rolling out AI tends to bring uncertainty, and people take their emotional cues from whoever's leading them. A leader who stays patient and cautiously optimistic sets a very different tone than one who reacts defensively. Picture a leader introducing a chatbot to handle routine support requests. Questions about job security pour in during the first few team meetings. The leader takes them slowly, without brushing them off, keeps the conversation focused on what's actually possible, shares a real plan for upskilling, and doesn't rush to fill every silence. That kind of composure calms the room and signals that learning matters more than blind compliance (Ramakrishnan & Krupskyi, 2025). Teams tend to mirror that steadiness, which makes the whole transition easier.

Motivation in Innovation

Motivating a team through digital change starts with showing real belief in both your people and the new tool. Leaders who learn alongside their team, instead of just directing from above, get more genuine participation. A department head running an "AI Hack Day," where the team explores new analytics tools on a real problem, and joining the sessions themselves, sharing their own mistakes along the way, does more for morale than a memo ever could. Over time, people stop being afraid to fail and start pitching bigger ideas. Motivation here comes less from cheerleading and more from a leader actually modeling risk-taking and curiosity in front of the team (Dwivedi, 2025).

Social Skills for Digital Communication

In a remote or hybrid team, relationship-building takes more intention. Without hallway run-ins or body language to fall back on, every message carries more weight than it would in person. A manager leading a distributed team through a new AI workflow might schedule regular video check-ins, send a note when a milestone lands, and open an anonymous channel for feedback. In meetings, they call on the quieter people directly and take a beat to clarify tone before a hard topic gets discussed. That extra attention keeps misunderstandings from piling up and keeps collaboration intact, and it tells the team their leader is actually reachable (Ramakrishnan & Krupskyi, 2025).

Empathy-Driven Communication Strategies

As more of daily work runs through AI, from automated workflows to data-driven performance reviews, team members increasingly interact with a system instead of a person. Clear, empathetic communication becomes critical here, not optional, for keeping morale and cohesion intact (Snyders, 2025). Leaders build confidence in this environment by being transparent, inclusive, and responsive on purpose.

Transparency around AI-driven decisions matters most when the stakes feel personal. When an algorithm has a say in who gets picked for a project or how goals are set, silence breeds suspicion fast. Good leaders explain, in plain language, how the tool actually reached its conclusion. A director might open an all-hands with a simple graphic showing how the project allocation tool weighs skill sets against recent workload. In an email update, a manager can explain the reasoning behind an outcome directly: "Our scheduling AI proposed this configuration based on availability and past collaboration patterns." Leaving room for questions after an announcement like that signals real openness, not just a courtesy (Snyders, 2025; Empathetic AI Policy Example: A Framework for the Human Impact of AI, 2025).

Active listening still matters, it just has more channels to work through now. AI-enabled surveys and suggestion boxes gather input efficiently, but leaning on them alone risks people feeling unheard. The leaders who get this right pair the numbers with a real conversation. A VP who notices a dip in a pulse survey and follows up with informal one-on-ones is reading both the data and what's underneath it, and making sure nothing important gets missed (Empathetic AI Policy Example: A Framework for the Human Impact of AI, 2025).

Inclusion needs extra attention once technology is standing between people. AI models can carry hidden bias, and remote settings can amplify some voices while quieting others. Leaders who take this seriously audit their AI tools for fairness, checking that an automated performance review isn't quietly favoring one group over another. They also make deliberate room for quieter people in meetings, sometimes by rotating who leads the discussion. If a manager notices a tool's recommendations rarely include cross-functional team members, that's worth questioning directly, whether that means adjusting the process internally or raising it with the vendor (Snyders, 2025; Empathetic AI Policy Example: A Framework for the Human Impact of AI, 2025).

Continuous feedback loops give a team both permission and a mechanism to raise concerns early. A feedback liaison who checks in regularly and brings suggestions to leadership meetings is one way to build that in. When leadership visibly acts on what they hear, adjusting a clunky AI dashboard after people flag it, for instance, that's what actually builds trust, more than the survey itself ever could (Empathetic AI Policy Example: A Framework for the Human Impact of AI, 2025).

Responding quickly to concerns matters most during a transition, when uncertainty is already high. A leader who thanks people for speaking up and follows through with specific fixes reinforces that it's safe to keep speaking up. If several people raise concerns about how a system tracks productivity, pulling together a focused session to clarify the policy and co-design some boundaries closes the loop in a way that keeps people engaged instead of quietly resentful (Snyders, 2025).

Taken together, transparency, real listening, deliberate inclusion, and a working feedback loop are what let empathy shape the daily relationships on a team, not just the official policy (Snyders, 2025; Empathetic AI Policy Example: A Framework for the Human Impact of AI, 2025).

Building Trust in AI-Enabled Teams

Open communication sets the tone, but real trust comes from what leaders actually do once AI starts influencing everyday decisions. When people see their manager being honest about what a tool can and can't do, skepticism turns into curiosity instead of quiet resistance. Trust here comes down to transparency, respect, and follow-through.

AI is genuinely useful, and it's not infallible. Leaders earn credibility by saying plainly what a new tool can do and where it falls short. A sales manager rolling out a predictive lead-scoring system might tell the team upfront that it sometimes misses context, a longstanding relationship, a recent change at the client. Something like: "This model helps us spot patterns, but you know your accounts better than any machine does. If something feels off, say so." Naming the limits up front sets realistic expectations and makes room for the team to actually push back when it matters (Empathetic AI Policy Example: A Framework for the Human Impact of AI, 2025).

Being upfront about where AI falls short does more than manage expectations, it teaches the team to think critically instead of treating the tool's output as gospel. A recruiter who notices a resume-screening tool consistently downgrading candidates with unconventional backgrounds should feel safe raising it, and does, when leadership has made clear that kind of scrutiny is welcome.

Real partnership with AI needs room for skepticism. The best leaders make it genuinely safe to question a recommendation without fear of looking difficult. One supervisor has the team review a weekly AI analytics report together, looking specifically for anything that seems off, and publicly credits whoever catches something. Another runs periodic "AI challenge" sessions, testing the system's logic against recent real cases, framed explicitly as learning, not blame. Either way, the message is the same: raising doubts is valued over quiet acceptance (Empathetic AI Policy Example: A Framework for the Human Impact of AI, 2025).

When people see their feedback actually lead to a fix, they start seeing themselves as part of building the system, not just using it.

Over time, that habit of shared scrutiny catches errors earlier and makes it normal to talk about mistakes instead of hiding them.

Clarity at the outset matters just as much. A project lead rolling out a task-scheduling assistant who explains up front what's automated, what stays manual, and where people can override the tool removes a lot of uncertainty before it starts. During onboarding, explaining what problem the system solves, how it's monitored, and who to contact with concerns does the same. Some teams write simple ground rules together, like "if the AI's recommendation doesn't make sense, flag it before acting." All of this helps people understand exactly where their own judgment still applies (Empathetic AI Policy Example: A Framework for the Human Impact of AI, 2025; The Human Side of AI Transformation: Why Culture Is the Key to Enterprise AI Success - Agility at Scale, 2025).

Open dashboards and plain-language briefings make it easier for people to see what's actually happening behind the scenes, which reinforces that the AI is there to support them, not replace them. Laying out a clear escalation path, when and how to bring in a human, builds a safety net into the everyday workflow and cuts down on unpleasant surprises.

AI can speed up analysis and take tedious work off someone's plate, but the actual breakthroughs still come from people bringing creativity and judgment to the table. Making a point of celebrating those moments keeps anyone from feeling reduced to a score. After automating part of customer support, a manager might share a story about the employee who caught an error the system missed, or call out a moment where someone's read on a situation prevented a real misunderstanding. Small rituals like that, a regular shout-out for a good catch or a well-handled judgment call, keep the message clear: AI is a tool the team uses, not a replacement for the team (Empathetic AI Policy Example: A Framework for the Human Impact of AI, 2025).

None of this is abstract. It shows up in honest conversations about trade-offs, in policies that actually reward people for speaking up, and in the everyday balance between using new technology and still paying attention to its impact on people. Leaders who talk about the tech honestly and back it up with visible investment in their team earn something durable: a group that's willing to experiment because they trust the ground they're standing on (The Human Side of AI Transformation: Why Culture Is the Key to Enterprise AI Success - Agility at Scale, 2025).

Bringing It All Together

This chapter comes down to one idea: empathy is still at the center of good leadership, even as more of the work moves through screens and systems. The tools keep changing, dashboards, virtual meetings, chatbots, but the need for real connection doesn't. Empathy isn't a box to check. It's the active work of noticing what's going on with people and doing something about it, even across the distance that remote work and digital tools can create.

The leaders who handle this well aren't the ones who master every new platform first. They're the ones who use empathy to build real trust across every interaction they have, digital or otherwise. Listening closely, communicating honestly, and keeping the feedback loop open add up to something no app can replicate: a team where people feel genuinely seen. Build those habits and your team won't just survive the next wave of change. They'll be ready for it.

Four

Ethical Leadership in the Age of Algorithms

It takes 20 years to build a reputation and five minutes to ruin it.

Warren Buffett

IT'S EASY TO THINK of AI as an impartial tool, quietly crunching numbers without any human baggage. Machines don't have feelings or opinions. But that view misses something important: AI systems learn from data shaped by human choices and societal bias. When the information feeding an algorithm carries the marks of old inequalities or narrow viewpoints, the AI reflects those flaws back, and sometimes makes them worse. This chapter looks at why treating AI as neutral is a myth, and why believing it can lead leaders somewhere they don't want to go.

Ethical leadership isn't only about spotting problems. It's about actively guiding a team through the biases hiding in algorithmic decisions. This chapter covers how to recognize where bias creeps in, how to demand transparency from the tools you use, and how to build accountability that keeps fairness at the center. Understood well, this lets people leaders use technology as a force for inclusion instead of exclusion, and lead through the complexity of AI with real integrity.

Debunking the 'Neutral AI' Myth

Chapter 1 covered how adapting to AI means challenging assumptions, not just learning new tools. The idea that technology is always impartial is one of the assumptions worth challenging most. AI's reputation for neutrality comes from its speed, its knack for pattern recognition, and its lack of visible emotion. But underneath that, every AI system is built by humans, people with values, blind spots, and limits, which means it can't separate itself from the society that built it. The myth persists because technical output looks objective even when it's quietly repeating old prejudice. For a leader steering a team through fast-moving digital change, taking that at face value is naive, and it's risky.

The reason this matters: when an organization trusts a black-box decision without scrutiny, real damage follows. AI doesn't come from nowhere. Every model digests old decisions, historical company records, or population-scale behavior, and hands back recommendations that often just resemble what came before. Picture an AI screening job applicants using years of resumes from a male-dominated industry. Unless diverse perspectives were deliberately built into the data, the model will likely favor candidates who look like the ones who succeeded in the past (Team, 2025). That isn't progress. It's bias automated and relabeled as efficiency. Any gap or distortion already sitting in the data becomes a blind spot in the algorithm's logic, invisible until someone thinks to question it.

Training data is where unintentional bias usually creeps in. Data is never neutral. It reflects who got hired, who got promoted, and whose performance got praised or overlooked. Feed a model imbalanced or incomplete history and it will perpetuate that imbalance, sometimes magnifying it. If women or people of color were historically underrepresented in senior roles, a model trained on that history can "learn" to downgrade similar candidates going forward. When University of Washington researchers had three AI language models rank resumes in 2024, the models favored white-associated names in 85 percent of tests and Black-associated names in 9 percent (Wilson & Caliskan, 2024). The system isn't inventing new prejudice. It's just very good at passing forward whatever was already baked into its training data. The quality and diversity of that input data isn't a minor technical detail. It's a management responsibility that shapes fairness, opportunity, and the organization's reputation.

Real cases make this concrete. Amazon's recruiting tool was trained on ten years of resumes from a period when tech hiring skewed heavily male, and the algorithm ended up favoring men's resumes, deepening the gender gap it was trained on (DigitalOcean, 2024). ProPublica's 2016 analysis of COMPAS, a risk score used in courtrooms to predict future crime, found that Black defendants who did not go on to reoffend were almost twice as likely as white defendants to be labeled higher risk

(Angwin et al., 2016). The company behind the tool disputed that analysis, and the debate that followed showed how hard it is to agree on what a fair score looks like. In both cases, flawed data plus limited human oversight let bias slip through and get amplified. The cost isn't abstract: missed opportunities, lost trust, legal exposure, and talented people who quietly opt out of applying in the first place.

Once a leader accepts that AI can reflect and reinforce unfairness, blind trust in algorithms stops being defensible. Responsible leadership means scrutinizing how data gets collected, demanding transparency from vendors and technical teams, and insisting on regular review of what the AI is actually producing (DigitalOcean, 2024). When a model ranks candidates or predicts creditworthiness, the useful question isn't just what the algorithm decided. It's why. Making it normal to question AI output turns ethical oversight into everyone's job, not just something technologists or compliance officers worry about.

Any people leader can put a few concrete steps in place: run bias assessments on a regular cadence, build project teams with varied backgrounds so more scenarios get anticipated and tested, and put routines in place for monitoring AI decisions and following up on flagged anomalies (DigitalOcean, 2024). Insisting on transparency, welcoming healthy skepticism, and supporting ongoing learning signals that an organization takes both innovation and inclusion seriously. It means making space for dissent, treating fairness as something to actually measure, and giving up the comfort of easy answers.

Recognizing these hidden influences is only the start. The next step is building a formal ethical framework, one grounded in real principles, so that people and AI systems both reflect what the organization actually stands for.

Establishing an Ethical AI Framework

Chapter 1 covered how the people leader's role now requires balancing human and machine strengths. Doing that well takes more than technical competence. Managers are becoming stewards of ethical standards in a workplace increasingly shaped by algorithms. A reliable ethical AI framework isn't optional anymore. It's foundational to leading responsibly with this technology (A Guide to AI Governance by Modulos, n.d.; Austin, 2025).

Four principles anchor this kind of leadership: fairness, transparency, accountability, and privacy. Practiced consistently, each one moves an organization closer to ethical maturity (A Guide to AI Governance by Modulos, n.d.).

- Fairness means AI tools treat every employee and stakeholder equitably. If an algorithm screens job candidates, its criteria shouldn't quietly favor or disadvantage any group. Leaders can push for this with regular audits of the decision process and by bringing teams into reviewing the outcomes directly.

- Transparency means people actually understand how and why a decision got made. If an employee questions a performance review generated by AI, publishing the criteria behind it and giving them a real channel for feedback turns transparency from a value into an everyday habit.

- Accountability means leaders own the consequences of an AI-powered decision no matter who wrote the code behind it. When an HR tool introduces unexpected bias, pointing at engineering or an outside vendor isn't enough. Naming the issue openly, fixing it, and sharing what was learned is what actually builds trust, inside the company and out.

- Privacy means sensitive data stays protected and gets used only for what it was collected for, whether that means restricting access to certain data types or anonymizing it before analysis.

Codifying these values into guidelines is the next step. Clear guidelines turn gray areas into actionable standards, but they only work if people actually see them: distribute them during onboarding, run workshops where people can ask real questions, and keep them visible where the work actually happens. A company whose AI ethics policy sat unread until a costly bias incident forced a rushed overhaul is a case study in what happens when guidelines exist but nobody knows them. Regular review keeps a policy responsive to new risks instead of frozen in time (Austin, 2025).

A framework is only as strong as who helped build it. Leadership can't write this policy alone, from the executive suite or from IT. Bringing in ethicists, legal advisors, frontline staff, and even customers surfaces blind spots that would otherwise turn into reputational damage later. Real inclusion means listening to people outside the dominant perspective, not just adding names to a list. Rotating who sits on an ethics panel, or holding a town hall open to anyone, junior analyst or outside community member, catches things a closed leadership group would miss. One company narrowly avoided a PR disaster when a summer intern flagged a flaw in a recruitment AI's scoring system, a problem the rest of the team had missed simply because nobody else had the intern's vantage point (A Guide to AI Governance by Modulos, n.d.).

An environment where learning and dialogue keep going over time is what makes this stick. Training on real dilemmas helps people internalize the standard instead of just memorizing it. Regular cross-team meetings keep ethics an active priority instead of a box to check once a year. Anonymous feedback channels give people who'd otherwise stay quiet a way to flag something early. Think of this less like a one-time vaccine and more like a daily vitamin: small, consistent effort that protects the organization from the slow-burning problems that build up out of sight. Leaders who keep this culture alive give their teams real permission to ask hard questions before a small misstep turns into a real one.

None of this works if ethics and technology stay in separate lanes. A good ethical framework isn't bureaucratic overhead, it's what makes a team more resilient, fairer, and more trusted. Teams built on shared values handle setbacks better and get more creative with how they use AI in the first place. Strong governance now is what makes the next hard call, on data stewardship or technical risk, easier to get right.

Handling Data Responsibly and Securely

Principles only matter once they show up in daily practice, especially with the sensitive data that powers most AI tools. Ethical leadership depends on what actually happens at every touchpoint: what data your team collects, how they share it, and how they keep it safe. Good data stewardship isn't an IT checkbox. It's how an organization earns trust and builds resilience over time. Even the best protocols can't answer every gray area, which is where the next section picks up.

Think of personal information like cargo on a ship: you load only what you actually need, and you guard everything you do load. In practice that means collecting the minimum dataset that actually serves the business goal. Extra data feels harmless until it's another entry point for a breach. A manager deciding whether to keep years-old employee records might ask a simple question: do we still need this for a real purpose? If not, delete it. Data minimization isn't just a technical policy, it's a mindset that respects consent and shows restraint. Anonymizing identifiers in training data cuts the risk further. Leaders who model "less is more" send a clear signal that privacy matters as much as efficiency, and that choice tends to pay off in both public trust and regulatory compliance (9 Best Practices for AI Compliance and How to Implement Them, 2025).

Trust is currency. Customers, partners, and employees all care how their data gets used, especially once AI is involved. A company hiding behind a dense, legalese privacy notice invites suspicion. One that runs live Q&As or publishes plain guides explaining what's collected, why, and how it's used, builds confidence instead. Explaining the "why" behind a data practice, instead of burying it in fine print, creates real buy-in from the marketing team down to the end user (Gaba, 2024). Sometimes transparency is as simple as a briefing that walks staff through a change in how data gets handled, or sharing an anonymized case study of what worked and what didn't. When people see leadership being open about this, they raise concerns earlier instead of assuming something's being hidden.

Good policy means little without real protection behind it. Encryption, access controls, and audit trails work like locks, alarms, and cameras for an organization's most valuable information. But security is also a culture problem, not just a technology one: state-of-the-art encryption doesn't help if someone's using a weak password or falls for a phishing email. Regular tabletop drills, simulated incidents, test whether a team is actually ready and remind everyone that security isn't someone else's job. Continuous monitoring and predictive alerts can catch irregularities before they escalate (9 Best Practices for AI Compliance and How to Implement Them, 2025; Gaba, 2024). Spot checks and rotating access privileges help keep complacency from setting in. Safe data handling has to become a habit the whole team practices, not a policy that sits in a drawer.

Regulations move fast. GDPR, CCPA, HIPAA, and industry-specific rules keep evolving, and non-compliance costs more than a fine. Fines sting, but a damaged reputation and lost customer trust cost longer. The fix starts with a manager who doesn't just delegate compliance and forget about it: schedule regular reviews with legal and IT, then translate the key points into plain English for the team. Quick-reference guides and periodic refreshers keep the requirements front of mind even for people who aren't specialists. AI models need ongoing attention too, since data drift can push results out of bounds over time, which is why frequent audits and retraining matter (9 Best Practices for AI Compliance and How to Implement Them, 2025). Careful documentation and real engagement with domain experts and ethicists keeps the important things from getting missed. Compliance isn't just a legal hurdle. It's part of the actual business strategy.

Even with airtight processes, gray areas show up. Unusual cases and new risks mean no protocol answers every question on its own. The next section covers how to navigate those dilemmas when the rulebook runs out.

Navigating Ethical Gray Areas in Algorithmic Decision-Making

Ethical dilemmas in algorithm-driven leadership rarely come labeled right or wrong. Picture an HR manager using an AI tool to screen applicants: it's efficient, but a few colleagues start quietly wondering why fewer candidates from certain backgrounds are making it through. Moments like that show how fairness and accountability actually live in the gap between policy and practice. When values collide, speed against inclusivity, say, leaders end up questioning not just the decision but the yardstick used to judge it. Noticing that gap isn't a failure. It's a sign someone's actually thinking it through instead of settling for an easy answer (Kandasamy, 2024).

Leaders can start by paying attention to early warning signs: a stakeholder complaint, an unexpected pattern in the data, a shift in team morale, or just a nagging feeling that the numbers don't tell the whole story. Instead of rushing to a fix, the better move is to pause and ask the harder questions. What assumptions shaped this model? Who might be affected in a way nobody's considered? Did someone raise a concern that never got answered? Teaching a team to examine both the visible results and the hidden impact builds a fuller picture of what's actually happening (Rodgers et al., 2022).

A structured decision framework can help here, not as a rigid formula but more like a flashlight for corners that would otherwise stay dark. Picture a leader weighing whether to roll out a new scheduling AI. They'd start by gathering input from shift workers, supervisors, and union reps, some of whom will push back hard, and treat each perspective as adding real dimension to the picture. Next comes naming the actual values in tension: operational efficiency against worker flexibility, laid out plainly rather than swept under the rug. From there the team brainstorms real options, a pilot group, an opt-out, manual override settings, and weighs the short-term gains against the long-term cost for different groups with each one. The final call isn't about what pleases the most people. It's about what actually reflects the organization's stated principles and its commitment to fairness.

Documenting each step can feel tedious under deadline pressure, but it matters. A clear record lets someone else trace the logic later and holds the whole process up to scrutiny. That transparency is what makes the next similar dilemma easier: more context, better answers, less confusion (Rodgers et al., 2022). Recording the reasoning, not just the conclusion, helps a team trust both the outcome and the process that got them there.

Outside perspective matters too. Even a strong internal team develops blind spots, sometimes just from getting used to how things have always worked. Bringing in an ad hoc peer panel or an ethics committee can surface something the core team missed. Inviting people unaffiliated with the project, alongside technical experts and honest outsiders, to review an AI launch plan often catches issues insiders walked right past. Sometimes the value of that outside review is simply making the ethical question impossible to quietly ignore.

Decision tools help, but they don't replace judgment. Using a framework in a genuinely complex situation doesn't make the path simple, it's more like a flashlight in a dark room than a well-lit highway. As one leader put it after revisiting a contentious AI rollout: if getting to yes felt too easy, something probably got missed. Real leadership means balancing time pressure, competing priorities, and incomplete information, and it's the slow, deliberate work of sitting in that gray space that actually builds trust over time (Kandasamy, 2024).

Think back to a recent gray-area moment in your own team, a time you felt pulled in two directions and wished for a clearer map. Try the same approach here: pause at the discomfort, bring in more than one voice, and write down why you chose what you chose.

Final Thoughts

The idea of neutral AI is more myth than reality. Every system you implement is shaped by choices, the data you feed it and the way its decisions get reviewed and refined. Bias doesn't have to be intentional to do damage. It hides in old habits, historical data, and assumptions nobody thought to check. None of that is a reason for discouragement. It's a reason to lean into the part of leadership that questions, audits, and leads with both curiosity and care. That influence shows up everywhere, from the questions you ask your team to whether diverse voices actually get heard before a blind spot becomes a real problem.

Building fair, accountable AI starts with ordinary leadership habits: a strong ethical framework, real input from diverse perspectives, and staying alert to the gray areas as they show up. Fairness, transparency, accountability, and privacy only matter as living priorities, not slogans, and treating them that way is what makes both your team and the systems they use more trustworthy over time. That's the tool that actually holds up as automation keeps reshaping the way people work.

Five

Strategic AI Implementation: From Vision to Practice

By far, the greatest danger of Artificial Intelligence is that people conclude too early that they understand it

Eliezer Yudkowsky

IT'S EASY TO GET caught up in the excitement of launching an AI initiative, energized by the promise of innovation and a real edge over competitors. Yet plenty of organizations end up stuck, tripped up by unclear goals and pushback from teams who aren't sure what AI actually means for their work. Projects stall, resources get wasted, and initial enthusiasm curdles into frustration. These aren't rare, unlucky roadblocks. They're common enough to derail even a well-intentioned effort if a leader doesn't address them from the start.

Crafting an AI-Driven Vision

Leading an organization into AI starts the same way any ambitious effort does: by defining what success actually looks like before plotting the path there. Leaders need clear, measurable business objectives before they get into the specifics of any tool. Instead of a vague aim like "make operations better," aim for something concrete: improve guest satisfaction scores by 15 percent, or cut operating costs by 5 percent. Measurable goals tie every AI initiative to what actually matters for growth (Develop Your Organization's Vision for AI, 2025; How to Align AI Development with Business Goals Using AI Strategy, 2025). Strategic alignment keeps a team's attention on the enterprise's actual priorities instead of whatever tech trend is loudest this quarter.

It's easy to get swept up in the promise of a breakthrough, but overpromising erodes trust fast. Offer realistic projections and don't frame AI as a magic fix. If the pitch sounds too good to be true, people will hesitate to invest real effort. The CLEAR framework, Collaborative, Limited, Emotional, Appreciable, and Refinable, is a useful way to set goals that are pragmatic and shared rather than imposed from above. It keeps morale and credibility intact while still allowing for continuous improvement (How to Align AI Development with Business Goals Using AI Strategy, 2025).

AI should never overshadow an organization's mission and values. It should amplify them. Picture the company's mission as a compass that always points north, with AI initiatives as new tools and shortcuts, not detours. Map each project so it fits into the existing workflow rather than disrupting it. If your company has a real track record on inclusivity, start by automating the processes that benefit employees broadly.

Reinforcing those principles in every AI decision keeps the team motivated and gives everyone a shared sense of purpose (Develop Your Organization's Vision for AI, 2025).

A vision works best as both shield and catalyst: it protects against confusion and fear while also sparking action. Communication is the center of that. When introducing AI, name the real anxieties out loud: job security, workflow changes, whether people still matter. Explain that AI is meant to lighten the manual load and make room for more meaningful work, not replace the person doing it. Use plain language, a metaphor that actually lands, and real stories from inside the company. A manager sharing a story about an AI assistant freeing up time for a customer care rep to actually solve a complicated client problem does more than any slide deck. Using more than one channel, email, video, a town hall, gives everyone a chance to actually receive the message and respond to it (How to Align AI Development with Business Goals Using AI Strategy, 2025).

Building buy-in isn't about broadcasting from the top. It's closer to tending a garden together. Bring in cross-functional input early through workshops, brainstorms, or informal listening sessions. One finance director asked colleagues from IT, HR, and marketing to help co-create the AI vision, and got to consensus faster with fewer surprises down the line. These sessions tap into insight that's otherwise scattered across departments. Equip your internal champions to model openness, relay feedback up to leadership, translate between technical and non-technical teams, and make sure early wins get celebrated loudly. Actually following through on what people tell you is what turns suspicion into trust (Develop Your Organization's Vision for AI, 2025; How to Align AI Development with Business Goals Using AI Strategy, 2025).

None of this works as a solo act. Build inclusion and ownership into every stage and you end up with a coalition durable enough to weather real setbacks. The vision stops being a slogan at that point and starts being something people actually use to make decisions.

Selecting the Right AI Tools

A clear, people-first vision only carries a leader so far. What comes next is making grounded choices about the actual technology. Every tool, dashboard, and integration should serve a real organizational goal, not chase whatever's trending.

Understanding Available Solutions

The first challenge is cutting through the noise to find tools that actually match your organization's pain points. One team might need an AI assistant for customer service, another automation for routine HR tasks. The key is mapping the requirement back to the broader vision: if the strategy calls for faster answers to employee questions, look for something that integrates with your existing communication channels. A mid-sized healthcare firm trying to cut down on manual data entry errors picked its AI invoice tool only after confirming it actually fit their regulatory workflow, a good model for how this should go.

Actionable takeaway: start with a checklist of must-haves tied to your vision. For each candidate tool, ask: which real business problem does this solve? How well does it fit existing processes? Can it scale as needs grow? That clarity keeps you from getting distracted by features that look impressive but don't add value. Tie the choice back to collaborative intelligence: what teams do best, supported by what AI automates most efficiently (Guidelines for Selecting and Using AI Tools in Growing Businesses | Xantrion, 2025).

Prioritizing User-Friendliness

However advanced a tool is under the hood, adoption comes down to usability. Can employees across roles and skill levels actually navigate the new platform with confidence? Is the interface intuitive? Do the dashboards surface what matters without burying it in clutter? Picture two AI project management apps rolling out side by side: one dazzles with complex analytics but confuses the staff using it, the other guides people step by step and keeps customization simple. The simpler one wins adoption fast, even among the skeptics, because it lowers the barrier to entry.

Worth asking directly: will people need extensive training? Does it allow real personalization for how teams actually work? Does it integrate cleanly with what they already use, Teams,

Slack, whatever it is? Prioritizing usability isn't just about productivity. It builds trust and gives everyone a real stake in whether this succeeds (Kumar, 2024).

Vendor Evaluation Criteria

Choosing an AI vendor deserves the same scrutiny as hiring a key team member. This is a long-term partnership, not a one-time purchase. Managers need to check providers for reliability, transparency, and fit with the organization's actual values. Data security comes first, especially with sensitive information involved: look for real certifications like ISO 27001 or SOC 2, solid encryption, and clear data-handling policies. Reliability matters too, uptime history, responsive support, evidence the vendor actually ships updates. Plenty of leaders have picked the budget-friendly option only to discover hidden costs once usage scales or a premium feature turns out to be essential. Read the contract carefully before that happens to you.

Worth a checklist here too: does the vendor provide real documentation? Is there an active support community? What's the actual escalation process if something breaks? These questions build the accountability that any real digital partnership needs (Kumar, 2024; Guidelines for Selecting and Using AI Tools in Growing Businesses | Xantrion, 2025).

Piloting Before Committing

Even with solid research, a tool's real behavior often only shows up once you're actually using it. Small pilots lower the risk, surface problems you didn't anticipate, and build buy-in early. A people leader might run a chatbot pilot in one department with clear metrics attached, faster response times, better employee satisfaction scores, whatever fits. Along the way, ask the people using it directly: was the guidance clear? Did the new process speed things up or just add friction? Were privacy concerns actually addressed?

A pilot also tells you something about cultural readiness. Resistance might mean people need more training. Enthusiasm might mean you've found a quick win worth scaling. Worth tracking as you go: what data will actually prove success? Who owns troubleshooting? How fast can you iterate on what you learn? A well-run pilot isn't a side experiment. It's a real step toward the bigger rollout (Guidelines for Selecting and Using AI Tools in Growing Businesses | Xantrion, 2025).

Choosing the right tools means moving past the demos and the buzzwords. Every decision should support real performance, real inclusivity, and real trust. Root the selection in organizational purpose, weigh usability seriously, vet vendors like partners, and pilot before committing, and you've built solid ground for what comes next: turning the plan into an actual rollout.

Building the Implementation Roadmap

Once the tools are chosen, the work shifts from planning to action. A roadmap is what actually gets a team from potential to progress. Without one, even a strong toolset can stall out in indecision or a pilot that never quite lands. Choosing the technology isn't the finish line. It's where translating vision into daily results actually begins (AI Readiness: Key Factors & Best Practices, 2025).

Quick wins set the tone. Instead of a massive, high-profile rollout, the smarter move is narrowing to a scenario with fast payback and low risk: a customer service team automating ticket triage, or HR using an AI assistant to answer routine questions about internal policies. These early projects should hit three marks: visible value, limited scope, and data that's actually accessible. Teams that see early value get more invested, and the fast feedback surfaces both the strengths and the blind spots in the broader strategy (IBM, 2024).

Milestones and metrics turn a big ambition into something trackable. Every phase needs clear deliverables, deadlines, and a way to measure progress, faster response times, fewer manual hours, model accuracy, adoption rates, whatever fits the project. A typical rollout might start with a small pilot, expand to a single region, then scale company-wide once it's proven itself. Checking in every couple weeks or monthly, depending on how fast the project is moving, catches problems before they snowball and shows leaders exactly where the friction points are, whether that's a skill gap, an unclear expectation, or a process that's just slower than expected (IBM, 2024; AI Readiness: Key Factors & Best Practices, 2025).

Resourcing this well means more than throwing money or extra licenses at the problem. Leaders need to name project leads, identify where skills are actually missing, and plan upskilling around that gap. Picture a busy analytics lead who ends up as the point person for a deployment, backed by scheduled training for the junior analysts on the team. Sometimes the right call is bringing in a machine learning specialist for a critical phase, paired with internal staff so the knowledge actually transfers instead of walking out the door when the contract ends. Budget for professional growth and cross-training, not just infrastructure, so teams feel real ownership instead of waiting around for outside expertise (AI Readiness: Key Factors & Best Practices, 2025).

Monitoring and adjusting is what keeps the momentum from stalling. Quarterly check-ins or after-action reviews give a team structured space to share what they've learned and raise honest concerns. A marketing team might only discover a compliance risk after a chatbot's already live, which means a mid-course correction. Some business units will run ahead with AI while others lag, and that's worth naming directly rather than ignoring. Adaptability matters more than sticking rigidly to the original schedule. A GPS recalculates the route after a wrong turn. A good roadmap should work the same way (IBM, 2024).

None of this happens in isolation. A roadmap draws on every decision that came before it: which tools got chosen, who's driving implementation, how a winning pilot scales across the organization. Get the milestones, the resourcing, and the flexibility to adjust all lined up, and a team learns fast, avoids repeating its own mistakes, and builds real trust along the way.

An adaptable, transparent plan does more than keep a project on track. It gives leaders something to navigate by when teams hit resistance or run into a wall of bad data, which is exactly what the next section covers.

Overcoming Barriers to Adoption

Even a leader who's built a solid vision and a real plan will run into practical headwinds once implementation is underway. Think of a relay team rounding the final curve: every runner has the baton, but one fumbled handoff can still change the outcome. Facing these barriers early, with clear eyes, is what lets a manager steer through them with confidence. Incremental milestones, open communication, and real buy-in aren't just administrative checkboxes. They're what makes it possible to weather the moments that actually matter (Caleb, 2024; Marr, 2024).

Resistance to Change

Few things cause more friction during a digital rollout than resistance to change. Picture announcing a new AI document review tool in a hybrid finance department: some people worry their expertise is being devalued, others feel shut out of the decision entirely. That's a normal reaction. It usually comes from fear of job loss, uncertainty about where they still add value, or plain skepticism toward anything new (Caleb, 2024; Marr, 2024).

The fix is straightforward, even if it takes patience. Host real Q&A sessions where leaders answer concerns honestly and explain how the tool supports the work rather than replacing the person doing it. Something like: "This platform helps us spot errors faster, so your judgment and experience get to do what they're actually good at." Regular updates and pilot tests that genuinely invite input reinforce that this is a shared effort, not something handed down. Publicly recognizing small wins, like a team that's adapted well to a new system, builds trust over time. This kind of transparency is the same thread running through the planning stages covered earlier.

When a mid-sized HR firm hit pushback over automated resume screening, leadership ran biweekly coffee chats: no slides, just honest conversation. Over time, employees didn't just accept the new process, they started suggesting other areas worth automating. The resistance turned into collaboration, which is exactly what patient, consistent communication tends to produce.

Skill Gaps

The second major barrier is skill gaps. A manager might be genuinely excited about a new AI-powered customer insights tool, only to realize as the deadline approaches that almost nobody on the team knows how to actually interpret the model's output, let alone tune it. That's a technical problem, but it's also a confidence problem (Caleb, 2024; Marr, 2024).

The fix goes beyond hiring. Upskilling programs, lunch-and-learn sessions, and university partnerships all build real expertise internally. For urgent needs, outsourcing complex modeling can close the immediate gap, as long as the external experts are paired with internal staff so the knowledge doesn't just walk out the door when the contract ends. Micro-credentials or public recognition for learning a new tool can turn skill-building into something closer to a team sport. None of this only closes a technical deficit. It builds a learning culture that makes the whole organization more ready for what comes next.

A retail chain closed its own analytics gap by running monthly "AI Showcases," where employees demo new techniques to each other. As participation grew, so did the chain's in-house capability, and data science stopped being seen as somebody else's job.

Data Quality Constraints

Data is to AI what soil is to a garden. Without something healthy underneath, nothing grows (Marr, 2024; Caleb, 2024). Halfway through a promising AI-targeted marketing campaign, a leader might discover duplicate records or missing details quietly sabotaging performance. Usually that's because data lives in silos, formats are inconsistent, or entry practices are outdated.

The fix: standardize data entry, clean the databases on a regular schedule, and put clear ownership in place. A data governance committee with both IT and business stakeholders keeps quality visible instead of forgotten. Modern integration tools make consolidating legacy data more reliable. Every improvement here reinforces the same trust-building work covered earlier, proving to a team that leadership actually cares about getting the data right.

One insurance company improved its claims-processing AI by training clerks to flag anomalies right at the point of entry instead of after the fact. Error rates dropped over the following months, and the team took real pride in the improvement, proof that better soil really does pay off.

Integration with Legacy Systems

Integrating AI with established systems is a bit like renovating a kitchen without disrupting family dinner. Outdated software, patchwork databases, and mismatched hardware can all slow the project down (Caleb, 2024; Marr, 2024).

The practical fix is gradualism. APIs and middleware let a company add AI capability incrementally instead of forcing a sudden, risky overhaul. Piloting small integrations in low-risk areas gives a team room to troubleshoot before scaling. Involving users in testing along the way builds familiarity and a real sense of ownership, the same inclusive approach covered earlier in this chapter.

A logistics provider started by linking AI-driven route optimization to a single regional hub, refining the connection before rolling it out nationwide. Iterating that way let them avoid a catastrophic failure and build confidence one region at a time.

Meeting these day-to-day obstacles, resistance, skill gaps, bad data, legacy systems, is what actually builds a leader's ability to support a team for the long haul. Working through them together is what makes the rest of the rollout possible.

Bringing It All Together

Steering an organization through AI adoption isn't about chasing trends. It's about setting a thoughtful course shaped by clear goals, real collaborative planning, and an honest eye on the actual challenges along the way. From a vision rooted in the company's values to tools that genuinely support the team, every step works best anchored in purpose instead of hype. Involving people early, listening to their concerns, and celebrating the small wins is what makes change feel like something built together rather than handed down from above.

No roadmap is perfect, and there will be bumps. But tackling resistance, closing skill gaps, cleaning up the data, and modernizing systems a step at a time turns those hurdles into real learning instead of eroded trust. Most of all, remember that a successful AI rollout was never just about efficiency. It's about giving people what they need to do their best work. What's covered here is the foundation. What comes next is putting it into practice, troubleshooting as you go, and leading with confidence through whatever shows up.

Six

Empowering Teams for the AI Era

Psychological safety is a sense of felt permission for candor.

Amy Edmondson

PICTURE A MID-SIZED TECHNOLOGY company standing at a crossroads, facing the same pressure every industry feels as AI reshapes it. Instead of scrambling or retreating, this company invested in its people: a company-wide AI upskilling initiative. Through steady emphasis on continuous learning, and a real cultural shift toward treating experimentation as something to celebrate rather than tolerate, employees built new skills that reshaped their actual day-to-day roles. This wasn't jargon-heavy, one-size-fits-all training. It was practical, hands-on, and tied directly to the problems people were already facing. The company didn't just survive the shift. It moved faster because of it, staying ahead of a market that wasn't waiting around.

Empowering Teams through Upskilling in the AI Era

Helping a team build durable skills for working with AI starts with figuring out what actually matters in your context. For a mid- to senior-level leader navigating this shift, that doesn't mean becoming a technical expert overnight. It means getting clear on what genuinely affects your team's day-to-day performance. Picture walking into a Monday morning meeting where your operations manager shares an AI-generated report flagging customer churn risk. The skill gap shows up immediately: interpreting that data and acting on it takes a different kind of analytical reasoning and business judgment than the job required five years ago. Working with your team to name the three most important AI-related capabilities for your context, maybe critical thinking, digital literacy, communicating well across remote settings, grounds the whole upskilling plan in something real (Tamayo et al., 2023; Mayer et al., 2025).

Actionable takeaway: map the three most essential AI-related skills your department actually needs, then have your team weigh in and validate the list.

Designing continuous learning into the workflow is what makes these skills achievable instead of aspirational. The old model of a one-off training session doesn't fit anymore. What works now is small, iterative learning that fits around a hybrid schedule. A product team might block off thirty minutes every

Wednesday for a quick session where two people share a new trick for using an AI-enhanced collaboration tool. Elsewhere, a video "lunch and learn" gives people a casual space to talk through recent struggles with automation software, share screens, and ask questions without feeling judged. These small sessions add up fast, folding learning into the actual flow of work instead of bolting it on. Recognizing the people who help colleagues troubleshoot a new platform or make sense of a confusing model output reinforces peer mentoring as a real, valued behavior.

Actionable takeaway: pilot a recurring microlearning session or a peer-led lunch discussion to get fresh knowledge circulating during the actual work week.

Bringing in outside learning partners adds expertise and energy a team can't always generate on its own. A local university partnership can mean an in-depth workshop or a tailored certificate program, ethical AI use is a common topic. One retailer facing hesitancy among frontline workers brought in an online learning provider for evening short courses, and employees could earn digital badges that gave their profiles, and their morale, a real boost. Another company partnered with a university to co-design project labs where teams solved simulated business problems using generative AI tools. Vendor resources are worth using too when rolling out a new enterprise system, quick-start guides, live Q&As, and troubleshooting chats often come bundled in and speed up how fast people get comfortable.

Actionable takeaway: reach out to at least one external learning partner this quarter and see what their offerings could add to your team's development.

Measuring the impact of all this takes more than asking whether people enjoyed the course. A good leader tracks participation, watches for real shifts in performance, and ties both back to business goals. When a services team leader noticed only a handful of people had finished the foundational AI workshops, digging into the attendance data revealed that compliance staff had opted out entirely, missing updates directly relevant to risk modeling. Tailoring the next round of training with direct input from the compliance leads fixed that. Another organization started tracking certification completion against operational metrics and found that teams with higher engagement in upskilling improved client response times measurably. Sharing results like that transparently is what secures buy-in from skeptical leadership and builds momentum for the next investment (Mayer et al., 2025).

Actionable takeaway: review your team's current learning metrics, look for participation gaps, and test new ways to tie skill-building directly to outcomes that matter for your strategy.

Investing in continuous learning now does more than meet today's demands. It builds the skills a team needs to experiment boldly later, which is where this chapter goes next. Strong skills are the runway teams need to take calculated risks and adapt under real uncertainty. Equip people now, and you're setting them up to actually use AI, not as passive consumers of a tool but as curious, capable partners in the work.

Fostering a Culture of Experimentation

As a team's readiness for AI grows through skill-building, the next challenge is turning that new capability into real innovation. Learning something and actually applying it are two different things, and closing that gap means deliberately building a culture that champions experimentation. Empower a team to try new approaches and you unlock their problem-solving instincts while also opening the door to a wider range of perspectives on what comes next. Building that culture takes a leader modeling openness, treating setbacks as something to learn from, and rewarding curiosity in ways people can actually see (Lee, 2025).

One of the strongest tools a leader has is normalizing trying, and failing. The taboo around failure still runs deep in a lot of organizations, and it kills risk-taking before it starts. A manager who wants real breakthroughs has to treat even a flawed experiment as a win for learning. Picture a director whose AI workflow pilot doesn't move the productivity needle the way she expected. Instead of quietly shelving the result, she runs a debrief with the team: where the initial assumptions went wrong, what data gaps showed up, what to try differently next time. Being open about her own missteps, and asking her team to do the same, reframes the disappointment as progress. The team learns that trying something imperfectly is still worth doing, because every attempt teaches something useful for the next one (Lee, 2025).

Psychological safety is what turns that attitude into a daily habit instead of a one-time speech. The belief that anyone can speak up without fear of embarrassment or punishment isn't a nice extra, it's the foundation experimentation actually rests on (Duhigg, 2016). For a leader, that means running meetings where every person genuinely has room to ask a question or float an unusual idea without getting shut down. Regular huddles and asynchronous forums both work as venues for this. The managers who do it well listen without rushing to judge and thank the people who bring up a problem others were avoiding. A manager who says "let's make our mistakes loud so we all learn faster" is signaling that surfacing uncertainty is a contribution, not a weakness, and that builds the trust that speeds up learning for everyone (Duhigg, 2016).

Recognizing curiosity keeps this momentum going. In organizations where innovation is more marathon than sprint, both formal and informal recognition matter. A leader might spotlight an "experimenter of the month" or hold a quarterly lunch celebrating the most inventive flops alongside the lucky breaks. Seeing a peer get recognized for a smart risk, a chatbot integration that initially flopped but surfaced a genuinely useful workflow insight, makes people more willing to test their own hunches. The best managers go beyond applause: extra time for a skill-building project, a digital badge, a spot on a high-visibility assignment tied to curiosity-driven work (Lee, 2025). These signal that the value isn't only in the win. It's in the willingness to keep chasing an idea.

Documenting what's learned keeps the cycle from resetting after every experiment. A leader should push for a real process, recording both what worked and what didn't so the knowledge actually accumulates. A shared digital knowledge base, project goals, outcomes, hurdles, and surprises included, becomes something new hires can learn from and something that sparks ideas across teams (Lee, 2025). After-action reviews and retrospectives give people a blame-free space to reflect honestly, which makes it easier to write down what actually worked and why. That institutional memory is what keeps a team from repeating its own mistakes and lets good ideas travel between departments instead of staying stuck in one.

Experimentation isn't just a step between upskilling and inclusion. It's what turns a technical skill into real confidence, and what lets every team member's contribution actually land. A leader who visibly embraces experimentation, builds psychological safety, rewards curiosity consistently, and makes sure hard-won lessons get shared is setting the stage for people from every background to bring something new to the table. In a workplace reshaped by AI, that's what lets a team adapt fast, keep learning, and catch opportunities no algorithm would have flagged (Lee, 2025; Duhigg, 2016).

Promoting Inclusion and Diversity: Empowering Teams for the AI Era

Diverse perspectives aren't a nice-to-have on an AI project. They're necessary for building systems that are actually ethical, effective, and resilient. When a team brings together people from different backgrounds, each person contributes something shaped by their role, their expertise, and their own lived experience (Shams et al., 2023).

Digital transformation moves fast, and resilience matters here. A team with a real mix of professional and cultural perspectives adapts faster because it's drawing from a wider range of experience. When regulations or consumer behavior shift suddenly, a genuinely cross-functional team can adjust strategy on the fly by pulling on that range. One global retailer's AI ethics group included supply chain managers from Asia, developers from Europe, and HR leaders from North America, and that mix is what led them to a real breakthrough in flagging biased hiring recommendations (Shams et al., 2023).

Bringing a broad set of voices into every stage of an AI project also matters for catching bias before it does damage. Left unchecked, bias in AI can quietly work its way into automated decisions in ways that are hard to see from the inside (Cachat-Rosset & Klarsfeld, 2023). When a team actually reflects the diversity of the people the system serves, hidden risks surface faster. A stakeholder workshop where frontline employees flag that a virtual assistant's language choices favor certain dialects is one example. A healthcare provider piloting a new diagnostic tool who brings in feedback from both patients and nurses, and catches a subtle gender bias in the symptom prompts as a result, is another.

Regular consultation sessions, open forums, and user testing with underrepresented groups make a real difference here. Bias-awareness training helps too, giving everyone the tools to recognize the "uninvited guests" that creep into AI systems and to see how their own assumptions shape a design decision, tied back to the organization's actual commitments on equity (Shams et al., 2023). Building inclusivity into everyday workflow, rather than treating it as a separate initiative, produces AI that genuinely serves both the workforce and the customer base better.

AI can do more than avoid harm. It can expand access, especially when accessibility is part of the project's core goals from the start. Automated captioning helps employees who are deaf or hard of hearing communicate clearly. Adaptive interfaces support people with visual, motor, or cognitive differences. When product managers bring in early feedback from users with disabilities, the resulting improvements tend to help everyone, the same way voice commands ended up helping anyone juggling multiple things at once, at work or at home.

A mid-sized tech firm used AI-powered translation to close the language gap between its offices across five countries. People who used to struggle across the language barrier started sharing ideas more freely, which sped up onboarding for new hires, including neurodiverse employees, and raised engagement across the board (Cachat-Rosset & Klarsfeld, 2023).

None of this holds up without tracking and real accountability behind it. Reporting on representation within AI project teams, who's actually participating in training, workshops, and pilots, surfaces gaps and shows real progress. A simple dashboard showing project roles broken out by gender, ethnicity, or function can catch an imbalance before it affects an outcome. Transparency also means checking the demographic impact of each AI solution on a regular basis. If a scheduling algorithm is quietly reducing advancement opportunities for part-time workers, a clear metric should catch that early enough to actually respond.

Tying managerial evaluations or incentive plans to real diversity and inclusion targets keeps this accountable at the top, an executive whose bonus is partly tied to progress on equitable access has a real stake in it. Simple, regular reporting, quarterly staff updates or an anonymized case study, shows a team that leadership is serious about this and invites everyone into the ongoing work (Shams et al., 2023; Cachat-Rosset & Klarsfeld, 2023).

Chapter 8 picks this up further, looking at how to sustain a high-performing, inclusive organization over time. What's covered here is the groundwork.

Coaching for Agile Adaptation

A team facing constant AI-driven change needs more than technical know-how. It needs the confidence to shift direction, learn in real time, and stay motivated as priorities move underneath them. Building that kind of adaptability is the natural next step for a leader who's already worked on inclusion, psychological safety, and a real culture of experimentation. This is where a team learns not just to welcome new ideas but to actually hold steady inside constant uncertainty.

Peer coaching circles are a practical way to build this into the routine. Picture a small group meeting every other Wednesday morning, coffee or video call, where each person brings a recent challenge: confusion over an updated analytics platform, a shifting deadline that threw off someone's workflow. The group listens, gives honest feedback, and works the problem together. Rotating who facilitates and keeping a couple of simple ground rules, confidentiality, equal airtime, turns these into a real space to learn from both wins and failures. In a hybrid environment where hallway conversations don't happen naturally, structured sessions like this keep trust and momentum alive (Steegh et al., 2024). A consistent rhythm, weekly or monthly, keeps things from backsliding and holds everyone accountable, which matters even more when the team is spread across time zones.

Adaptability doesn't come naturally to everyone, so it helps to teach it directly through agile frameworks that turn uncertainty into something workable. Think of agile methods less like a paved highway and more like whitewater rafting: you prepare for the turbulence by staying alert, paddling together, and adjusting course quickly. A daily stand-up, a quick round where each person names yesterday's progress, today's goal, and any blockers, surfaces problems early. A retrospective at the end of each sprint gives the team a moment to reflect together: what worked, what needs adjusting, what new risk showed up now that AI tools are part of the process (Sureshkumar Somanathan, 2023). A shared board where distributed teammates log issues and share updates between meetings keeps everyone looking at the same picture even when they're not in the same room.

Leading through this kind of change relies less on technical expertise and more on setting an honest, steady tone. When something big shifts, an AI-based resource allocation tool going live, say, a leader can ease the anxiety by explaining not just what's changing but why. Picture a director running an open Q&A about a new system: taking the hard questions ("What happens if we miss something?"), admitting there will be bumps, actually inviting suggestions. When a team member raises a concern about losing oversight, the director doesn't brush it off: "That's a fair point. Let's agree on some early signals to watch, so nobody feels left in the dark," followed up after the meeting with an actual plan. That kind of calm, visible honesty tells a team there's room for a mistake and a course correction, and it roots resilience in the relationship rather than in a rigid policy (Steegh et al., 2024). When something does go sideways, a leader who stays composed and open shows the team the setback is temporary and solvable, not a crisis.

Even in an agile team, a leader can't lose sight of people's actual well-being. AI-driven change speeds things up, but it can also fuel burnout if workloads creep unchecked or ambiguity goes unaddressed (Sureshkumar Somanathan, 2023). A regular check-in, a five-minute survey or a short call just to ask "how's everyone actually doing," catches this early. If a leader notices fatigue creeping into status updates during a demanding rollout, offering optional time off or trimming the meeting load keeps engagement sustainable instead of burning people out. Naming and redistributing an unfair load before it does damage protects morale and shows real care without sacrificing output. That means being clear about workload expectations, offering flexibility where it's possible, making mental health resources genuinely accessible, including virtual ones, and giving people real permission to say no when they're at capacity.

Bringing It All Together

Empowering a team in the AI era has less to do with mastering every technical detail and more to do with building a culture where learning and curiosity actually drive the daily work. Focusing on practical skills like analytical thinking and digital literacy, and bringing the whole team into that conversation, is what sets a team up for smarter decisions and faster innovation.

As the technology keeps changing, the leaders who build real inclusion, welcome new ideas, and support agile adaptation will find their teams don't just get through it, they get stronger because of it. Small, consistent moves, regular peer coaching, celebrating an honest failure, building meetings where everyone actually gets heard, are what build the trust and resilience that hold up under real pressure. It starts with one step: encouraging your team to learn, adapt, and experiment together, and that's genuinely where the real progress comes from.

Seven
AI-Driven Decision-Making and the Role of the Manager

We demand rigidly defined areas of doubt and uncertainty!

Douglas Adams (The Hitchhiker's Guide to the Galaxy)

PICTURE A SEASONED MANAGER deep in a crucial decision, leaning on years of intuition built through countless scenarios. At the same time, an AI system hums in the background, crunching vast datasets to suggest the next move. One relies on gut feeling shaped by experience and subtle cues. The other parses patterns invisible to human senses, offering recommendations built from cold data and fast calculation. That's the tension running through this whole chapter: the old wisdom of people leadership meeting the new capability of machine intelligence.

Human vs. AI Decision-Making: Strengths and Weaknesses

Picture a regional sales manager staring at a dashboard of red, green, and amber signals minutes before quarterly numbers get reported. Her gut says something's off with one account's performance, but the analytics platform is already generating recommendations pulled from hundreds of thousands of trends across similar organizations. That's the moment human judgment and AI's raw processing power collide, both aiming at the same outcome from two different directions.

AI's biggest advantage is sheer speed and scale. Reviewing every transaction for a new product launch by hand means one person working through a spreadsheet. An AI system parses millions of data points in seconds, flagging unexpected returns or subtle patterns a person would never catch by eye. That reach lets AI spot risk and opportunity far beyond what any individual analyst could manage alone, which matters enormously in time-pressured moments like a quarterly close, where slowness has a real cost. Speed here doesn't just mean a faster dashboard. It means catching a market shift before it snowballs into a crisis. But volume of analysis isn't the same as accuracy. Sometimes the signal that actually matters drowns in statistical noise, or a real exception gets lost inside the relentless logic of the code.

Data alone rarely captures the full reality inside an organization. Experienced leaders know that key decisions hinge on context and nuance, the exact things an algorithm struggles to read. Take a merger: AI might suggest optimizing reporting lines based on historical integration success rates, but it won't catch the tension simmering between two teams or notice that one department feels blindsided by the news. A manager tuned into the office undercurrents might hear the hesitation in someone's voice on a call, or read the room during a hallway conversation, and those small signals often matter more than anything an algorithm can aggregate. Human judgment pulls together ethics, cultural fit, reputation risk, and what stakeholders actually care about, especially when the path forward isn't written anywhere in the historical data.

However sophisticated, AI approaches ambiguity limited by its training data and whatever boundaries a person set on its learning. People, by contrast, are good at making value-laden calls with incomplete information. Picture a team conflict after a disappointing product sprint: an automated suggestion about productivity tweaks will never substitute for a manager's read on what actually broke the trust, or how two clashing personalities are dragging down group cohesion. Managers calibrate that kind of judgment from lived experience and instinct, a skill built from hard lessons, not written in code. When reputational risk or an ethical gray area shows up, the advantage tips even further toward a human call.

Routine, repetitive, heavily quantitative work is where AI does its best work. Automating supplier reorders when stock runs low fits it naturally, and so does monitoring real-time competitor pricing shifts that would overwhelm a small analyst team. Clear rules, high volume, that's AI's territory. But the moment the landscape turns unpredictable, expanding into a new market during regulatory upheaval, handling a PR crisis after a viral story, the game changes entirely. Human flexibility and the ability to hold messy, conflicting inputs together is what actually helps here. If a major client suddenly threatens to walk, no model can capture years of partnership history or the informal commitments made over lunch.

The sharpest managers learn to recognize which kind of challenge they're actually facing. A clear-cut, number-heavy process? That's AI's sweet spot. Organizational turbulence, fast-moving politics, ambiguous goals? Instinct and experience need to lead. Sometimes the two blend: an AI engine might flag that engagement scores predict turnover, but only a manager who knows the actual people behind those numbers can decide whether the fix is flexible scheduling, public recognition, or rethinking the team structure entirely.

The real payoff comes when both strengths combine. Picture a retail chain using AI to anticipate seasonal swings in foot traffic, then letting store managers adjust staffing based on local events or community sentiment the model can't see. Or a talent review where the analytics flag skills gaps, but it's the manager's read on mentorship dynamics that actually shapes the promotion call. In both cases, a human is filtering the machine's output through values and lived experience, which is what keeps the result grounded in real-world complexity. That combination is what avoids groupthink and catches disruptive threats that don't match any historic pattern, while still moving fast enough to matter.

Managers running today's hybrid, tech-forward teams face both sides of this. They need to know what each tool actually offers, and when to switch from data-driven logic to empathetic judgment and back. Mastering that switch is what keeps an organization agile and resilient enough to turn uncertainty into an actual advantage.

Data-Driven Leadership: Best Practices

Getting a dashboard full of charts is one thing. Knowing what to actually do with it is another. Plenty of leaders end up staring at rows of numbers and visualizations without a clear read on how those numbers should change a decision that affects customers, teams, or the bottom line. Data-driven leadership isn't about flooding your calendar with reports. It's about turning insight into a real action that creates value.

Data literacy is the starting point. Leading well here doesn't require becoming a statistician, it requires asking the right questions and recognizing a few key concepts: averages, anomalies, trends. A sales manager reviewing a forecast that shows a sudden spike, with basic literacy, knows to ask whether that's a seasonal sale or an outlier customer before panicking into an over-order. One VP caught a real error just by asking how "average order value" was actually calculated, and found a handful of unusually large corporate deals skewing the whole number. Building this is straightforward: sit in on regular analyst briefings where questions are actually welcomed, attend a short internal workshop, or run through a checklist before accepting an AI-generated recommendation at face value. These habits make data part of the normal conversation instead of something handed down as dogma on a slide (Tableau, 2023).

Working well with data experts unlocks even more. Instead of treating analytics as a separate step tacked on at the end, bring analysts and business leaders together early and keep them together. Meet regularly to review not just the results but the logic behind the key metrics, and co-design the indicators together so they reflect what actually matters, not just what's easy to measure. One director partnering with a data scientist during a product launch defined the leading and lagging indicators that would genuinely signal success, which let the team pivot fast from vanity metrics to ones that actually mattered as conditions shifted. A cross-functional task force mixing marketing managers and data specialists is well suited to a stubborn problem like churn or engagement, since it forces the solution to consider more than one angle. Clear communication and real mutual respect are what close the gap between technical rigor and business pragmatism, and that's what keeps teams aligned instead of talking past each other (Gleeson, 2025).

Sharing insight well turns raw numbers into a story people actually rally around. A dashboard full of bar charts clarifies a trend, but the real work is connecting that chart to why it matters. A people leader presenting turnover rates might use a simple line graph, then ground it in a specific story about how an onboarding fix reduced exits among new hires. That's what makes the stakes feel real. Anchor each meeting around two key visuals, and coach team leads on how to actually weave a data point into a daily priority instead of just displaying it. Pairing a number with something qualitative, an employee comment, a customer review, a specific win, turns abstract analysis into something people can act on (Tableau, 2023).

Healthy debate keeps an organization sharp. Accept every insight without question and blind spots pile up quietly while groupthink takes over. Build an environment where poking holes in an assumption is not just safe but expected. A monthly "data challenge" session where people propose alternative explanations works well, and so does assigning a rotating devil's advocate whose job is to hunt for gaps or unexplored angles before a big decision gets made. One VP who gathered feedback from both frontline and executive staff before adopting an AI-recommended hiring process ended up modifying it after stakeholders raised a real bias concern, avoiding a misstep that could have hurt both diversity and morale. Leaders who model that kind of curiosity show the team that real debate is a strength, not a threat (Gleeson, 2025; Tableau, 2023).

Once these habits are part of the daily routine, decisions get faster and more confident. Asking clarifying questions, running joint KPI workshops, pairing data with a real story, building space for honest dissent, these are what let a manager at any level actually act on information instead of just collecting it. A high-trust, high-curiosity culture like this gets steadily better at handling ambiguity, which matters more every year as AI plays a bigger role in business strategy (Gleeson, 2025; Tableau, 2023).

Avoiding Decision-Making Pitfalls

Data literacy and real debate between managers and analysts matter, but even with both in place, common mistakes still catch leaders who lean too hard on AI-driven insight. This next level of decision-making calls for vigilance, not just confidence.

The first pitfall: overtrusting the algorithm. Picture a hiring manager handed a shortlist from an impressive new recruitment platform. The recommendations look solid, so the manager skips their own review and rubber-stamps the interview invitations. A month later, team dynamics are off, some of the new hires aren't landing well, and the fit feels wrong. Why? The algorithm was trained on limited historical data, and it carried old preferences forward into roles they didn't actually fit anymore (Roller, 2023). Trusting an AI recommendation without scrutiny is like following a GPS straight into a lake. It's giving guidance, but you're still the one steering. Every automated output deserves a real sense check: does this actually make sense in our context? What's the logic underneath it? Who set the criteria inside that black box? A little curiosity and healthy skepticism keeps judgment sharp.

Confirmation bias is another hazard on this same road. A leader who focuses only on the metrics that match their gut can miss a real warning sign. Picture a launch where sales look strong in some segments while early churn signals show up quietly in others, it's tempting to highlight the good dashboard and skip past the trouble. Biased AI can make this worse by serving up data already filtered through its own lens. To prevent that tunnel vision, it helps to actively hunt for disconfirming evidence: invite an opposing view into the planning session, or bring in an outside reviewer to audit the analysis. Diverse perspectives break the cycle of self-reinforcement and force a team to face an uncomfortable truth before it gets worse. Rotating who reviews the data, or opening an anonymous feedback channel, can surface a blind spot nobody knew was there.

Ignoring outliers and edge cases is the next trap, the small stuff that sometimes signals real trouble ahead. If an operational dashboard flags a handful of urgent complaints out of thousands of satisfied customers, it's tempting to write them off as noise. But like miners once watched canaries for early warning signs, a wise manager treats an outlier as a signal worth investigating. Those few upset customers might point to a systemic flaw nobody's noticed yet, or a gap a competitor could exploit. Instead of brushing it aside, a strong leader digs in: why did this happen? Do we actually understand the root cause? That curiosity is what turns a minor alarm into a chance to course-correct before it becomes expensive.

Human accountability is the most important checkpoint in all of this. Algorithms getting smarter doesn't shift responsibility away from people. If a marketing strategy falls flat or a key hire fails badly, "the algorithm decided" won't satisfy a stakeholder, a regulator, or a customer (Roller, 2023). A leader who deploys a cost-cutting AI tool only to watch morale and productivity drop can't explain that away in the boardroom by saying the software made the call. Documenting decision points, when and why algorithmic advice was followed or overruled, keeps this transparent, builds internal trust, and prepares the organization for whatever audit or investor question comes later. When in doubt, choose thoughtful deliberation over automated convenience. AI should be a trusted advisor here, not something to hide behind when a decision goes wrong.

Scenario Planning with AI Tools

The habits covered so far, continuous learning, curiosity, real experimentation, are exactly what a manager needs to use predictive analytics, simulation, and stress testing well in day-to-day work. Here's how those tools actually make a decision more informed and resilient.

Understanding Predictive Analytics

Predictive analytics uses AI models to turn both historical data and new information into a real forecast. Think of it as a weather app for your business: instead of guessing tomorrow's forecast off a gut feeling or an aching joint, you're using decades of climate data plus today's air pressure to get a much sharper read. In practice, these models can analyze sales trends, predict seasonal workforce needs, or anticipate churn (Bailey, 2015).

Unlike a static chart or a monthly report, AI-powered forecasting keeps refreshing. As new events unfold, a sudden regulatory change, a viral social media moment, the model absorbs the input and updates its predictions, surfacing an emerging threat or opportunity fast. A manager who actually uses this well can move from reacting after the fact to leading from ahead of it, steering the team proactively instead of playing catch-up (Sheremeta, 2026).

To put this into practice, identify the specific metrics where an early prediction would actually change a decision. Start small, a staffing forecast, a revenue projection, and compare the AI's read against your own judgment side by side. That comparison is what builds real trust in the process without handing over full control.

Simulating 'What If' Scenarios

AI-powered simulation brings scenario planning to life. Instead of debating a strategy around a table, a team can now test each option virtually before making a real commitment. Picture running the numbers on launching a new product line next quarter versus waiting, or seeing what happens if supply chain costs spike overnight. These simulations show the ripple effects of each choice, rewards and risks both, with a clarity a meeting rarely produces (Bailey, 2015).

A finance team can simulate thousands of economic scenarios in minutes, factoring in interest rates, raw material costs, or shifting consumer demand. The same applies operationally: HR might model headcount changes across markets, or a hybrid workplace manager might compare resource allocation strategies to find the mix with the most flexibility and the least downside. When a simulation exposes a hidden cost or an overlooked gap, it saves a leader from a costly trial-and-error mistake in the real world (Sheremeta, 2026).

The practical move: build simulation into the regular planning cycle. Run a few options through the tool and let that feedback sit alongside the group discussion, not replace it. Treat a strategy that fails in simulation as a real lesson, not a dead end.

Stress-Testing Assumptions

AI takes scenario analysis a level deeper by rigorously stress-testing an assumption from every angle, catching a weakness even a seasoned veteran might miss. Consider a technology rollout plan where a human review misses a rare but serious supplier risk, while an AI model processing enormous historical data flags it instantly (Sheremeta, 2026). Catching that early turns a near-miss into an actual success. This kind of digital stress test builds real confidence, not because it guarantees a good outcome, but because it prepares a team for a wider range of realities.

Stress tests aren't about predicting disaster. They're about building resilience. Regularly subjecting a plan to a simulated shock, a currency swing, a cyberattack, a sudden regulatory change, is how an organization gets better at spotting and fixing its own weak points. Schedule these tests routinely, treat a bad result as something to learn from rather than something to hide, and bring cross-functional teams in to interpret the results together. That's what makes finding a vulnerability feel like smart risk management instead of a hunt for someone to blame.

Making Decisions Under Uncertainty

Even with all these tools, uncertainty never fully disappears. AI offers real guidance on the likelihood of different outcomes, but it can't account for every variable or eliminate every surprise. A manager still has to balance a clear data signal against their own experience, the team's read on a situation, and the organization's actual values.

Treat AI's output like a trusted advisor: it sharpens thinking and offers real perspective, but the manager stays the pilot. Blend the probabilistic forecast with your own contextual read, and let each one challenge the other (Bailey, 2015; Sheremeta, 2026). When you feel stuck, try asking: what would I do if this probability chart disappeared tomorrow? That question is often what keeps a leader agile and decisive even when the path ahead needs both a sharp model and a steady hand.

Bringing It All Together

Human intuition and AI-driven analysis aren't rivals fighting for control of the decision. They're two different strengths, and the real skill is knowing which one the moment calls for. Seasoned judgment, workplace awareness, and ethical reasoning give a leader an edge exactly when a situation turns messy or uncharted. AI brings scale, speed, and the ability to sift through more information than any person could hold in their head, catching a pattern a person would miss entirely. The real challenge for a manager now is sensing which tool, gut feel or data, the moment actually demands, and knowing when to bring both together for something stronger than either alone.

Putting this into practice means building the habit of asking questions, checking the context, and welcoming honest debate around every real decision. A leader who takes both the dashboard's unexpected signal and the team's subtle cue seriously ends up with a fuller picture and usually a better call. As workplaces keep getting more digital, the leaders who stay curious, keep people at the center, and stay open to learning will be the ones best positioned to turn complexity into something they can actually use. Leading in the AI era was never about giving up control. It's about steering with both hands still firmly on the wheel.

Eight

Sustaining High-Performing, Inclusive Organizations Through AI

I'm sorry, Dave. I'm afraid I can't do that.

HAL 9000

PICTURE WALKING INTO AN office where AI does more than automate tasks. It actively helps a team reach further than it could alone. Data-driven insight surfaces what needs to happen and how to do it better and faster. Employees aren't just clocking hours, they're recognized for the specific strengths they actually bring. Learning happens daily, shaped by tools that adapt to how each person actually works, and diversity shows up as a real, visible thing in fair, clear performance measures rather than a word on a values poster. Leadership here isn't about controlling technology or people. It's about guiding real collaboration between the two, so fresh ideas actually have room to happen.

Redefining High Performance in the AI Era

Leaders working through digital transformation are realizing that tracking activity alone doesn't reveal progress. AI is good at surfacing which tasks actually move a strategic goal forward. For a sales team, the number of calls made used to be the metric. Now customer lifetime value or account expansion tells the real story. Leaders should get their teams asking which results actually matter to the mission, and build KPIs around outcomes instead of activity. When everyone's aiming at the same target, energy and focus multiply, and engagement follows.

What's driving this shift is a new generation of KPIs, built by algorithms that catch subtle patterns and hidden drivers of success. These smart KPIs don't just count. They learn. Companies using machine learning have found real performance markers that legacy metrics were missing entirely (Schrage et al., 2024). That pushes a leader past scorekeeping and into actual discovery. The teams that come out ahead will be the ones using smarter, adaptive KPIs that adjust as markets shift and surface an advantage the old measures never would have caught (Rashid & Kausik, 2024). Practical takeaway: review your current KPIs. Do they capture something that actually matters, or are they leftovers from an old routine? Push your team to build at least one genuinely outcome-oriented KPI every quarter.

Continuous learning is table stakes for a high performer now. Skill sets shift fast in this era, and adaptability is worth more than any fixed expertise. Recognize and reward the willingness to keep learning, not just the credential someone already has. An employee who actually uses a company-sponsored upskilling course is better positioned to pivot when the next tool shows up. Regular AI-powered skill assessments can flag a growth area or a hidden strength instantly, turning development into an ongoing process instead of an annual event. Build in peer recognition for the people helping others learn, and set the tone yourself by celebrating a real moment of growth or a recovery from a setback.

Personalized development is already here, quietly reshaping how people grow. AI-driven training adjusts to what someone actually needs, when they need it. A project manager might get a tailored microlearning module right after a tough assignment. A team lead's coaching plan might evolve based on recent performance data and strengths the system picked up on (Schrage et al., 2024). When a development plan actually mirrors someone's ambition, they feel valued, and they stay. Using AI recommendations to make feedback land personally, rather than generically, naturally makes room for different backgrounds too, since it's built around the individual instead of a template.

Innovation used to sit in its own department. Now it's something the whole team owns. The best organizations reward a real creative risk, not just the safe bet. A team might get real recognition for a pilot that didn't hit its numbers, because they documented what they learned and shared it openly. A quarterly innovation retrospective, or a small incentive for an unconventional fix, signals that trying something is what actually fuels progress. Compare that to an organization stuck in a blame cycle, where fear kills an idea before it ever gets tried. Reframe failure as learning, and teams get bolder, which is a real competitive edge in an industry that's moving fast (Rashid & Kausik, 2024).

Redefining high performance in the AI era, at its core, means shifting from busy to actually effective: real impact, real growth, development that fits the person, and innovation that isn't afraid to fail. A leader can start today by auditing an outdated KPI, building in room for adaptable learning, and championing creativity at every level. Each of those makes a team more ready for what's coming.

Making Inclusion Central to Digital Transformation

Inclusion isn't a moral checkbox on a digital transformation project. It's what sustained performance, resilience, and trust actually depend on. Once AI is woven into the daily workflow, raw output stops being the standard that matters most. Outcomes do, especially the ones tied to learning, adaptability, and innovation, and a team that feels genuinely seen produces those outcomes consistently (Buonocore et al., 2024; Tagscherer & Carbon, 2023). Turning that from a talking point into reality means building inclusive intent into every phase of adopting new technology, which is what makes a team both more robust and better prepared for whatever comes next.

Inclusive Design Principles

Real inclusive transformation starts with who's actually in the room when something gets designed. A diverse design team widens the lens a problem gets solved through, and catches a blind spot a more homogeneous group would miss entirely. Pulling in people with different backgrounds, disciplines, and lived experience makes it a lot less likely you'll accidentally build something that excludes the people who need it most.

Testing with a genuinely varied group matters just as much. Picture rolling out a workplace chatbot tested only by people working onsite: remote employees will hit problems fast, a login security step that doesn't account for offsite access, say. Testers pulled from a real cross-section of roles and situations catch that before launch, which cuts down on rework, builds real user confidence, and shows early that inclusion here is more than a talking point (Buonocore et al., 2024).

Monitoring Impact on Diversity

You can't manage what you don't measure. Understanding how a new technology actually affects diversity and equity takes real, data-driven evaluation. Take hiring: after rolling out an AI resume screener, a leader reviewing candidate advancement data broken out by gender, ethnicity, disability status, and other relevant categories might find underrepresented candidates consistently stalling out in the first round. Instead of treating that as a failure, the right move is convening a team to adjust the training data, tune the algorithm, and rework the prompts. Regular, transparent reporting lets staff track that progress themselves instead of taking leadership's word for it.

This kind of monitoring is what makes an agile feedback system possible in the first place. When people actually see leadership addressing a disparity in real time instead of talking around it, credibility grows. A team knows its well-being isn't resting on a vague promise. It's tied to something visible (Buonocore et al., 2024).

Adaptive Organizations: The New Competitive Edge

Diversity and innovation matter, but neither holds up without real adaptability underneath. Responding well to fast change, especially the kind AI keeps driving, is what lets an organization lead instead of just survive. Building that adaptability is what turns a reactive team into a genuinely proactive one.

Embracing uncertainty starts with accepting that market conditions don't pause for anyone to catch up. They move regardless. A leader who treats uncertainty like an open question instead of a threat creates room for real experimentation. In practice, that means highlighting what got learned when an experiment fails, not only when it succeeds. Make it a habit to ask what new idea got tried last week and what came of it. Small habits like that build the psychological safety that lets a team treat ambiguity as an opportunity instead of a liability (Chughtai et al., 2023).

Even the boldest idea stalls if every decision has to climb to the top of the org chart first. Flattening the hierarchy means trusting the people closest to a problem to actually act on it. One fast-moving product team swapped a sprawling approval chain for smaller, autonomous squads who could greenlight a change to a customer interaction using real-time data, cutting the bottleneck and giving people real initiative. That takes a leader delegating both authority and trust on purpose, training frontline staff to actually make the call, and recognizing an independent win when it happens. This isn't about tearing down structure entirely. It's building a path for information and action to move in both directions. Shared power tends to make an organization faster and more confident in how it responds to a real challenge (Sott & Bender, 2025).

A genuinely adaptable culture runs on responsive feedback, more than an annual survey ever provides. That means an ongoing back-and-forth, powered by both a person's read and AI-driven analytics. Short, regular check-ins where people can name a real hurdle, paired with a leader who actually follows up fast, build trust and speed at the same time. Give people an easy way to raise something, an app, a quick call, and then close the loop visibly by showing what actually changed because of it. Quick adaptation isn't the same as a knee-jerk reaction. It's a transparent, visible cycle of feedback and response. One simple habit: don't just listen, show people exactly how what they said shaped a decision. That visible responsiveness spreads on its own (Chughtai et al., 2023).

Investing in resilience is what makes any of this sustainable. Getting through one storm isn't enough. A leader has to prepare people to recover repeatedly. Think of resilience training as a fire drill: a regular, deliberate scenario that lets a team practice adjusting under pressure before a real crisis hits. That might mean simulating a supply chain breakdown, role-playing an unexpected departure, or just setting aside time to debrief after a big initiative, win or loss. On the well-being side, flexible schedules, peer mentoring, and real recognition for a personal win all support the recovery side of that equation. Build these into the routine and confidence, not panic, is what shows up when change actually hits.

Actionable tip: run a quarterly "what-if" session where the team imagines its response to a wild-card event. Over time that builds real organizational stamina and gets everyone more ready to meet something unexpected head-on (Sott & Bender, 2025).

These four things, embracing uncertainty, flattening hierarchy, building real feedback loops, and investing in resilience, reinforce each other. A team that's comfortable experimenting gets more out of a decentralized structure, and a resilient organization absorbs feedback better in the first place. Adaptive leadership was never about holding the spotlight. It's about equipping the people around you for whatever comes next, and building something that outlasts you (Chughtai et al., 2023; Sott & Bender, 2025).

Charting Your Leadership Legacy in the AI Revolution

Every theme in this book, human-AI partnership, empathy, ethical decision-making, eventually points to the same question: what will your leadership actually mean, not just this quarter, but years from now?

Self-reflection is where a lasting legacy actually starts. A lot of leaders build reflection directly into their routine: a weekly journal entry about a hard call, a quiet walk to replay a recent challenge. Setting aside time to ask what values you actually upheld this week, or whether your choices lined up with the organization's real purpose, surfaces something worth knowing. A tool like the Johari Window, or just asking for candid feedback directly, can turn up a blind spot and a moment worth being proud of, both (Meier, 2024).

Sharing what you've learned extends your impact well past your own growth. Mentoring someone means turning hard-won experience into something they can actually use. That doesn't need a formal program: a monthly coffee chat with a high-potential manager, or an informal roundtable where the team talks through what a recent AI project actually taught them, both count. As a mentor, push a mentee to pair technical reasoning with empathy and moral judgment, and remind them that real progress usually comes from wrestling honestly with fairness and what actually serves other people (Finlay, 2024).

Being honest about your own missteps, not just your wins, is often what actually sticks with someone years later.

Purpose-driven AI takes real intention, especially as automation reshapes more of the world by the day (Finlay, 2024). Building societal benefit into your KPIs and reporting, so people understand that the goal is for technology to lift the work up, not just optimize it, matters here. In practice that might mean adding equity outcomes to a performance dashboard, or running a cross-functional workshop that actually debates the ethics of a new tool honestly. Real leadership here was never about being first to market. It's about staying in the deliberate, sometimes messy conversation that actually builds trust.

Lifelong learning keeps a legacy from going stale. Staying curious keeps both you and your organization ready for what's next. A manager might spend an hour every other week on a webinar about data privacy, or a meetup on inclusive algorithm design, then bring the key takeaway back to the team and spark a real debate. Setting a quarterly goal to explore something unfamiliar, or teaching a new idea back to the team at the next all-hands, keeps that habit alive. The point was never collecting credentials. It's building an environment where growth doesn't stop, and where everyone feels like they're allowed to keep changing (Meier, 2024).

A culture built on reflection, mentorship, values-driven decisions, and real learning creates ripple effects that outlast any single quarter. Every small choice, the question you asked, the risk you reconsidered, the encouragement you actually gave, adds up. A tree planted today outlives the person who planted it, and a real leadership legacy works the same way, it extends past a quarterly result or even your own tenure. What actually lasts is the example: staying humble, sharing what you know, inviting real disagreement, and backing a team that brings its whole self to the work. As the technology keeps advancing, remember that a real breakthrough still comes from human intention, something no machine replicates on its own (Finlay, 2024; Meier, 2024).

Bringing It All Together

High performance and inclusion were never actually separate goals, and AI, used well, is what lets them reinforce each other. Focus on results that actually matter instead of busywork, build learning into the role itself instead of bolting it on, and let data support a genuinely fair, transparent decision, and a team gets somewhere it couldn't reach alone. The technology is just the starting point. The real advantage comes from using it to build a workplace where people feel seen, valued, and actually want to do their best work.

Staying adaptable is what keeps a team ready for whatever's next. Keep experimenting, keep the feedback flowing, and remember that real progress usually starts small, rethinking a KPI, naming a learning win out loud. AI was never here to replace what makes people valuable. It's here to back up what people are already good at, and open up new ways to work together. What you actually do with all of this, starting now, is what shapes not just a stronger team, but a more inclusive, resilient organization for years after this book is closed.

Conclusion

As you close this book, take a moment to appreciate the ground we've covered together. Leading people through uncertainty is never easy, and adding artificial intelligence to the mix can feel like being handed a Rubik's Cube with invisible colors. But you picked up this book because you saw opportunity instead of just challenge. You believed, as I do, that effective leadership in an AI-driven age isn't about being the most knowledgeable tech wiz in the room. It's about blending humanity and innovation to spark real change. My mission here has been to arm you with practical frameworks, relatable stories, and the confidence not just to survive this shift, but to actually shape it.

Throughout these chapters, we've returned again and again to one truth: technology's value isn't measured by its complexity. It's measured by how well it serves people. The leaders who leave the most meaningful legacy are the ones who pair empathy with curiosity and boldness with humility. No matter where you sit on your org chart, or how new you are to AI, you now have a toolkit for making sound judgments, building trust, and navigating disruption ethically and inclusively.

Let's step back and look at the bigger picture we've pieced together. At the center of modern leadership is a real, working partnership between humans and AI. You've seen how collaboration fuels creativity and problem-solving, how ethical leadership steers a team away from a costly misstep, and how an inclusive culture unlocks what a team is actually capable of. AI literacy, once just a buzz phrase, is now something you actually practice, whether you're translating technical jargon into plain English or questioning how an algorithm actually reached its decision. And adaptability isn't a soft skill anymore. It's the thing that lets you bend without breaking as the pace of change keeps accelerating.

Along the way, this book asked you to challenge some assumptions. AI augments your ability to inspire, coach, and connect, it doesn't replace it. Empathy in a digital workplace has to become more intentional, sometimes intentional enough to pause a chatbot rollout until you're sure your team actually feels heard. And keeping an ethical lens sharp was never a one-time checkbox. It's a daily habit, and it's the actual bedrock of resilient leadership.

If you're feeling both energized and a little daunted right now, that's a good sign. It takes real courage to sit with a topic that has no easy answer, admit what you don't know yet, and try anyway. Every chapter you finished was a small act of that kind of courage. You've wrestled with ambiguity, questioned your own assumptions, and chosen curiosity over comfort. That's what actually makes a leader who won't get sidelined by the next disruption, one who'll treat it as a chance to grow instead.

This book was only ever meant to be a start. AI, ethics, and team culture keep evolving, sometimes overnight. Treat this less like a manual and more like a passport, a way into further exploration, not the final stop. Go deeper into other resources, join a real conversation about AI in practice, host a lunch-and-learn on digital transformation, find a mentor or a mentee in your own network. Keep the learning going, and your team stays ready for what's actually next instead of just keeping up with what's already here.

Now comes the part that actually matters: action. Take what you've read here off the page and put it to work. Open up a real conversation about responsible AI use in your next team meeting. Start something small, an experiment in inclusive design, an honest ethical review, and see where it goes. When resistance shows up, and it will, go back to what this book covered on handling fear, closing a skill gap, and building real psychological safety. Don't wait for perfect clarity before you move. Progress beats perfection, every time.

Transformation multiplies when it's shared. Pass along what you've learned, including your own mistakes, to the people around you. Bring your peers into a real conversation about AI adoption. Celebrate your team's experiments even when they flop, because every honest attempt adds to what the whole team knows. It starts small: a single story, a question asked at the right moment, or just the willingness to say, "I don't have all the answers, so let's figure it out together."

One More Tool for the Road

If you want a concrete starting point, there's a short companion diagnostic built for exactly this book: The Decision Line, free, no account required, about five minutes. It walks you through ten workplace scenarios and gives you a real read on where your own delegation and transparency habits actually sit right now, plus which of four leadership tendencies you lean toward. It's a useful gut-check before you try to change anything, since it's hard to build a habit you can't see clearly yet.

decisionline.whitepine.pub

Above all, thank you. Thank you for putting your attention, your skepticism, and your hope into these pages. I'm genuinely grateful you trusted me as a guide through this complex, exhilarating shift toward responsible, human-centered AI leadership. My part in this doesn't end here. I want to hear your wins and your questions. Reach out, share what you're finding, and let's stay in the conversation. True leadership was always a dialogue, not a monologue, and this one's just getting started.

A Final Ask: Share Your Voice

If you've made it this far, thank you. That means you care deeply, not just about keeping up with change, but about leading through it with integrity, empathy, and curiosity. You've wrestled with the possibilities and pitfalls of AI, reflected on your own role as a leader, and hopefully walked away with a few new strategies and mindsets to bring back to your team.

Now I have a small request: would you take a moment to leave a review?

Your voice matters more than you might think. A review does more than help this book reach other readers. It signals that leadership in the age of AI isn't just a tech topic. It's about people. It's about trust, courage, and collaboration. Your perspective might be exactly what someone else needs to feel ready to get started.

Whether it's a few sentences or a full reflection, your honest feedback helps keep this conversation going. What resonated with you? What made you pause and think? What would you want another manager or leader to know?

If this book helped you navigate a tough conversation, rethink your approach to automation, or just feel a little more confident about what's next, say so. If something didn't work for you, that's just as valuable. I'm learning too.

You can leave your review wherever you bought the book, and it would genuinely mean a lot.

Let's make sure the future of work is smart and human, both.

Thank you for reading, and for being part of this conversation.

References

9 best practices for AI compliance and how to implement them. (2025). Sendbird. https://sendbird.com/blog/ai-compliance-best-practices

Amane Hoshimi. (2025, July 16). Digital Coworkers: Designing Workflows Around Human-AI Collaboration. Itbusinesstoday. https://itbusinesstoday.com/tech/ai/digital-coworkers-designing-workflows-around-human-ai-collaboration/

Angwin, J., Larson, J., Mattu, S., & Kirchner, L. (2016, May 23). Machine Bias. ProPublica. https://www.propublica.org/article/machine-bias-risk-assessments-in-criminal-sentencing

AI Readiness: Key Factors & Best Practices. (2025, June 27). Nexla. https://nexla.com/ai-readiness/

AI is not the silver bullet: How smart leaders drive lasting transformation. (2025). Potentialproject.com. https://www.potentialproject.com/insights/ai-is-not-the-silver-bullet-how-smart-leaders-drive-lasting-transformation

Austin, Thomas. (2025). Ethical AI Leadership: A Practical Guide for Managers Navigating Responsible Technology Implementation, Amazon.com.

https://www.amazon.com/Ethical-Leadership-Navigating-Responsible-Implementation-ebook/dp/B0DYYMZCRC

A Guide to AI Governance by Modulos. (n.d.). Modulos. https://www.modulos.ai/guide-to-ai-governance/

Buonocore, F., Annosi, M. C., Gennaro, D. de, & Riemma, F. (2024, September 27). Digital transformation and social change: Leadership strategies for responsible innovation. Journal of Engineering and Technology Management; Elsevier BV. https://doi.org/10.1016/j.jengtecman.2024.101843

Byrne, E. (2023, March 29). AI in the workplace: 12 ways workplace AI can improve the office. OfficeSpace Software. https://www.officespacesoftware.com/blog/ai-in-the-workplace-ai/

Bailey, K. (2015). How AI Transforms Scenario Analysis in Corporate Finance. Corporate Finance Institute. https://corporatefinanceinstitute.com/resources/fpa/ai-scenario-analysis-corporate-finance/

Caleb, A. (2024, July 3). Overcoming Challenges in AI Adoption. Researchgate. https://www.researchgate.net/publication/382108051_Overcoming_Challenges_in_AI_Adoption

Cachat-Rosset, G., & Klarsfeld, A. (2023, February 22). Diversity, Equity, and Inclusion in Artificial Intelligence: An Evaluation of Guidelines. Applied Artificial Intelligence. https://doi.org/10.1080/08839514.2023.2176618

Chughtai, M. S., Syed, F., Naseer, S., & Chinchilla, N. (2023, April 27). Role of adaptive leadership in learning organizations to boost organizational innovations with change self-efficacy. Current Psychology; NCBI. https://doi.org/10.1007/s12144-023-04669-z

Develop Your Organization's Vision for AI. (2025). Trailhead. https://trailhead.salesforce.com/content/learn/modules/ai-strategy/develop-your-organizations-vision-for-ai

Dwivedi, D. (2025, January 14). Emotional Intelligence and Artificial Intelligence Integration Strategies for Leadership Excellence. Advances in Research; Sciencedomain International. https://doi.org/10.9734/air/2025/v26i11235

Duhigg, C. (2016, February 25). What google learned from its quest to build the perfect team. The New York Times. https://www.nytimes.com/2016/02/28/magazine/what-google-learned-from-its-quest-to-build-the-perfect-team.html

DigitalOcean. (2024). Addressing AI Bias: Real-World Challenges and How to Solve Them | DigitalOcean. Digitalocean.com. https://www.digitalocean.com/resources/articles/ai-bias

Eklund, A. B. (2024, October 9). Balancing Humanity with AI in Leadership: A Leadership Coach's Perspective. Www.aipmo.org. https://www.aipmo.org/balancing-humanity-with-ai-in-leadership-a-leadership-coachs-perspective/

Empathy: What It Means for an AI-Driven Organization. (2025, February 26). Workday Blog. https://blog.workday.com/en-us/empathy-what-it-means-for-an-ai-driven-organization.html

Eastwood, B. (2025, February 3). When humans and AI work best together — and when each is better alone | MIT Sloan. MIT Sloan.

https://mitsloan.mit.edu/ideas-made-to-matter/when-humans-and-ai-work-best-together-and-when-each-better-alone

Empathetic AI Policy Example: A Framework for the Human Impact of AI. (2025, May 30). Solutions Review Technology News and Vendor Reviews. https://solutionsreview.com/empathetic-ai-policy-example-a-framework-for-the-human-impact-on-ai/

Frimpong, V. (2025, April 9). The Impact of AI on Evolving Leadership Theories and Practices. Deleted Journal. https://doi.org/10.53935/jomw.v2024i4.1100

Finlay, V. (2024, September 9). AI and Moral Reasoning in Leadership. The HOW Institute for Society. https://thehowinstitute.org/ai-and-moral-reasoning-in-leadership/

Gaba, N. (2024, May 23). How is AI Transforming Data Security Compliance in 2024? Compunnel. https://www.compunnel.com/blogs/the-intersection-of-ai-and-data-security-compliance-in-2024/

Gleeson, B. (2025, March 12). 5 Ways To Transform Leadership Through Data-Driven Decision-Making. Forbes. https://www.forbes.com/sites/brentgleeson/2025/03/12/5-ways-to-transform-leadership-through-data-driven-decision-making/

Goswami, R. (2025, June 29). Decision Making Models in AI Leadership: Are You Building Accountability on the Loop? - CTO Magazine. CTO Magazine. https://ctomagazine.com/ai-leadership-decision-making-models-accountability/

Goswami, R. (2024, August 30). Breaking AI Myths to Harness Its Power in the Modern Workforce - CTO Magazine. CTO Magazine. https://ctomagazine.com/ai-myths-workforce/

Guidelines For Selecting And Using AI Tools In Growing Businesses | Xantrion. (2025, February 28). Xantrion | IT Support and Cybersecurity Services | Bay Area. https://www.xantrion.com/blog/guidelines-for-selecting-and-using-ai-tools-in-growing-businesses

How to Align AI Development with Business Goals Using AI Strategy. (2025). Debutinfotech. https://www.debutinfotech.com/blog/align-ai-development-with-business-goals

IBM. (2024, November 21). Artificial intelligence implementation: 8 steps for success. Ibm.com. https://www.ibm.com/think/insights/artificial-intelligence-implementation

Kumar, V. (2024, July 16). Evaluating and Selecting the Right Generative AI Tools and Technologies. Association of Data Scientists. https://adasci.org/evaluating-and-selecting-the-right-generative-ai-tools-and-technologies/

Kandasamy, U. C. (2024). Ethical Leadership in the Age of AI: Challenges, Opportunities and Framework for Ethical Leadership. Arxiv.org. https://arxiv.org/html/2410.18095v2

Lewis, A. (2021, July 14). Soft Skills Training | Harvard Business Impact. Harvard Business Impact.

https://www.harvardbusiness.org/insight/5-key-human-skills-to-thrive-in-the-future-digital-workplace/

Lee, S. (2025). Mastering Learning Culture. Numberanalytics.com. https://www.numberanalytics.com/blog/mastering-learning-culture

Marr, B. (2024, May 10). 11 Barriers To Effective AI Adoption And How To Overcome Them. Forbes. https://www.forbes.com/sites/bernardmarr/2024/05/10/11-barriers-to-effective-ai-adoption-and-how-to-overcome-them/

Masood, A. (2025, June). AI in Organizational Change Management — Case Studies, Best Practices, Ethical Implications, and Future Technological Trajectories. Medium. https://medium.com/@adnanmasood/ai-in-organizational-change-management-case-studies-best-practices-ethical-implications-and-179be4ec2583

Mayer, H., Yee, L., Chui, M., & Roberts, R. (2025, January 28). Superagency in the workplace: Empowering people to unlock AI's full potential. McKinsey & Company. https://www.mckinsey.com/capabilities/mckinsey-digital/our-insights/superagency-in-the-workplace-empowering-people-to-unlock-ais-full-potential-at-work

Meier, J. (2024, December 29). How to Start Your Leadership Journey: A Step-by-Step Guide | JD Meier. JD Meier. https://jdmeier.com/how-to-start-your-leadership-journey/

Nikola Stanin. (2025, January 30). Redefining Leadership in the Age of Human-AI Teams: From Commanding to Orchestrating. Silicon Valley Innovation Center.

https://siliconvalley.center/blog/redefining-leadership-in-the-age-of-human-ai-teams-from-commanding-to-orchestrating

Palomino, I. (2025, February 22). Top 10 Qualities of Effective AI Leaders in the Tech Era. Peoplekult. https://www.peoplekult.com/post/top-10-qualities-of-effective-ai-leaders-in-the-tech-era

Ramakrishnan, V., & Krupskyi, O. (2025). EI & AI In Leadership and How It Can Affect Future Leaders. SSRN Electronic Journal. https://doi.org/10.2139/ssrn.5045017

Rashid, A. B., & Kausik, A. K. (2024, August 1). AI Revolutionizing Industries Worldwide: a Comprehensive Overview of Its Diverse Applications. Hybrid Advances; Elsevier BV. https://doi.org/10.1016/j.hybadv.2024.100277

Rodgers, W., Murray, J. M., Stefanidis, A., Degbey, W. Y., & Tarba, S. Y. (2022, June). An Artificial Intelligence Algorithmic Approach to Ethical decision-making in Human Resource Management Processes. Human Resource Management Review. https://doi.org/10.1016/j.hrmr.2022.100925

Roller, A. (2023, September 8). AI hiring bias: How HR can understand and mitigate potential pitfalls | HRMorning. HR Morning. https://www.hrmorning.com/articles/ai-hiring-bias-understand-and-mitigate/

Sott, M. K., & Bender, M. S. (2025, March). The Role of Adaptive Leadership in Times of Crisis: A Systematic Review and Conceptual Framework. Merits; Multidisciplinary Digital Publishing Institute. https://doi.org/10.3390/merits5010002

Shams, R. A., Zowghi, D., & Bano, M. (2023, November 13). AI and the quest for diversity and inclusion: a systematic literature review. AI and Ethics; Springer Nature. https://doi.org/10.1007/s43681-023-00362-w

Sureshkumar Somanathan. (2023, August 7). Artificial Intelligence Driven Agile Project Management: Enhancing Collaboration, Productivity, and Decision-Making in Virtual Teams. Nanotechnology Perceptions; Collegium Basilea. https://doi.org/10.5281/zenodo.15270083

Schrage, M., Kiron, D., Candelon, F., Khodabandeh, S., & Chu, M. (2024, February 12). The Future of Strategic Measurement: Enhancing KPIs With AI. MIT Sloan Management Review. https://sloanreview.mit.edu/projects/the-future-of-strategic-measurement-enhancing-kpis-with-ai/

Steegh, R., Van De Voorde, K., & Paauwe, J. (2024, October). Understanding how agile teams reach effectiveness: A systematic literature review to take stock and look forward. Human Resource Management Review. https://doi.org/10.1016/j.hrmr.2024.101056

Sheremeta, A. (2026, July 6). Preparing for the Unknown: Strategies in Scenario Modeling and Stress Testing. Dataforest.ai; Aleksandr Sheremeta. https://dataforest.ai/blog/scenario-modeling-and-stress-testing---the-art-of-navigating-uncertainty

Snyders, H. (2025, July 10). Empathy as a Change Management Tool for Your Organization. MSSBTA. https://www.mssbta.com/post/empathy-as-a-change-management-tool-for-your-organization

The Human Side of AI Transformation: Why Culture Is the Key to Enterprise AI Success - agility at scale. (2025, April 2). Agility at Scale. https://agility-at-scale.com/implementing/human-side-of-ai-transformation/

Tableau. (2023). *Data-driven decision making: Succeed in the digital era.* Tableau Software. https://www.tableau.com/learn/articles/data-driven-decision-making

Tagscherer, F., & Carbon, C.-C. (2023). Leadership for Successful digitalization: a Literature Review on Companies' Internal and External Aspects of Digitalization. *Sustainable Technology and Entrepreneurship;* Sciencedirect. https://www.sciencedirect.com/science/article/pii/S2773032823000032

Tamayo, J., Doumi, L., Goel, S., Kovács-Ondrejkovic, O., & Sadun, R. (2023, September 1). *Reskilling in the Age of AI.* Harvard Business Review; Harvard Business Review. https://hbr.org/2023/09/reskilling-in-the-age-of-ai

Team, E. (2025, May 28). *AI Bias: How It Impacts AI Systems [+Best Practices to Prevent It].* Tredence. https://www.tredence.com/blog/ai-bias

Uy, H. J. (2024, December 4). *How can empathy be the antidote to an overreliance on digital communication.* MIT Sloan Management Review Middle East. https://www.mitsloanme.com/article/how-can-empathy-be-the-antidote-to-an-overreliance-on-digital-communication/

Wilson, K., & Caliskan, A. (2024, October 16). Gender, Race, and Intersectional Bias in Resume Screening via Language Model Retrieval. Proceedings of the AAAI/ACM Conference on AI, Ethics, and Society. https://doi.org/10.1609/aies.v7i1.31748

Made in the USA
Monee, IL
04 November 2026

67378721R00075